Burning Earth: Unveiling the Flames

Steele Andrew Darren

Published by Steele Andrew Darren, 2024.

BURNING EARTH: UNVEILING THE FLAMES

First edition. March 15, 2024.

ISBN: 979-8224796236

Written by Steele Andrew Darren.

Table of Contents

Chapter 1: Introduction- Background on Global Warming and its Impact on the Planet

1.1 Rising concern: Global Warming's Shadow

Global warming has become an increasingly pressing issue in recent years, capturing the attention and concern of scientists, policymakers, and the general public worldwide. As our planet experiences rising temperatures, shifts in weather patterns, and a myriad of environmental upheavals, understanding the background and consequences of this phenomenon becomes crucial.

This chapter aims to provide an in-depth analysis of the historical background and scientific foundations of global warming. It will explore the impact of rising greenhouse gas emissions, the role of human activities in exacerbating this crisis, and the subsequent consequences on our planet.

1.2 Historical Context: From Ice Ages to the Greenhouse Effect

To fully comprehend global warming, it is essential to scrutinize climate change throughout history. The geological timescale has witnessed numerous fluctuations between ice ages and warmer periods, driven by natural factors such as volcanic activity, changes in solar radiation, and the composition of the atmosphere.

Since the industrial revolution, human beings have contributed substantially to altering this delicate balance through their actions. Jurij Brězan, a prominent climatologist, first introduced the concept of the greenhouse effect in 1824, elucidating the role of heat-trapping gases in the Earth's atmosphere. Scientists in the 20th century further developed this theory, fostering our understanding of the relationship between human activities, greenhouse gas emissions, and global temperature rise.

1.3 Gasping for Breath: The Acceleration of Global Warming

Over the past decades, global warming has advanced at an unprecedented rate, demonstrating alarming consequences for our planet, ecosystems, and human civilization. The emission of greenhouse gases, particularly carbon dioxide (CO_2), has skyrocketed due to the combustion of fossil fuels, deforestation, and industrial practices.

This acceleration has led to far-reaching impacts. Glaciers and ice sheets are rapidly melting, causing a rise in sea levels that threatens coastal communities. Extreme weather events, such as hurricanes and heatwaves, have become more frequent and intense. Shifts in precipitation patterns have resulted in droughts in some regions and floods in others. The intricate web of life on Earth is struggling, with many species facing extinction due to the destruction of habitats and shifts in ecosystems.

1.4 The Human Factor: Anthropogenic Global Warming

Scientific consensus overwhelmingly supports the notion that human influences are the primary drivers of global warming. The release of greenhouse gases from human activities, including burning fossil fuels, deforestation, and industrial processes, has disrupted the natural balance of the planet.

The growing world population, coupled with increasing consumption patterns and energy needs, has exacerbated these trends. Our dependence on fossil fuels, despite their finite nature, has amplified the climate crisis, as carbon dioxide levels surge to unparalleled levels in the atmosphere.

1.5 The Road Ahead: International Efforts and Mitigation Strategies

Recognizing the urgency, governments and organizations worldwide have taken steps to combat and mitigate global warming. The United Nations Framework Convention on Climate Change (UNFCCC) was established in 1992, and the subsequent Paris Agreement in 2015 set ambitious goals to limit global warming below 2 degrees Celsius above pre-industrial levels.

Mitigation strategies include transitioning to renewable energy sources, implementing energy-efficient technologies, promoting sustainable land use practices, and reducing greenhouse gas emissions. Adopting these measures at a global scale will be fundamental for ensuring the preservation of our planet's ecosystems and the wellbeing of future generations.

1.6 Chapter Overview

This chapter serves as a foundation for the subsequent discussions on the impacts of global warming. By understanding the scientific principles, historical context, and human contributions to this crisis, we can navigate the complex terrain of climate change with a well-informed approach. Embarking on this journey of knowledge will empower us to seek effective solutions, adapt to changing circumstances, and ultimately mitigate the adverse effects of global warming on our beloved planet.

- Introduction to the fossil fuel industry and its significance in the discussion

Of energy sources and climate change.
The fossil fuel industry plays a vital role in our current energy infrastructure, accounting for a significant portion of the world's energy consumption. Fossil fuels, which include coal, oil, and natural gas, have been essential sources of energy for centuries, revolutionizing human civilization and shaping the modern world as we know it.

One of the primary reasons for the significance of the fossil fuel industry is its abundance and accessibility. Fossil fuels are found in vast quantities across the globe, making them a readily available source of energy. This accessibility has allowed societies to harness and utilize these resources to power industries, transportation, and residential needs.

Furthermore, the energy density of fossil fuels is unmatched by alternative sources, providing a high-power density that cannot be easily replaced. This characteristic allows for efficient production and consumption of energy, helping to drive economic growth and development.

However, this reliance on fossil fuels comes with environmental and climate implications. Burning fossil fuels releases carbon dioxide and other greenhouse gases into the atmosphere, contributing to climate change. The combustion of coal, oil, and natural gas is the largest source of global greenhouse gas emissions, accounting for approximately three-quarters of total emissions.

These emissions trap heat in the atmosphere, leading to rising global temperatures, melting ice caps, increased frequency of extreme weather events, and sea-level rise. Climate change poses substantial threats to ecosystems, human health, and socio-economic systems worldwide.

As such, the role of fossil fuels in causing climate change has spurred discussions about alternative energy sources. The imperative to transition to more sustainable and cleaner energy options has gained traction and become a central facet of international efforts to combat climate change.

Renewable energy sources, such as solar, wind, hydro, and geothermal power, have emerged as potential substitutes for fossil fuels. These sources offer significant environmental benefits, as they produce little to no greenhouse gas emissions during operation. However, there are still challenges associated with their scalability, intermittency, and cost-effectiveness on a global scale.

Moreover, the fossil fuel industry has significant economic and geopolitical implications. Many countries heavily rely on fossil fuels for revenue generation through exports and employment opportunities. For instance, oil-rich nations in the Middle East have built their economies on oil production, creating a heavy dependence on fossil fuels.

The fossil fuel industry's economic importance is further underscored by the fact that many countries continue to heavily subsidize fossil fuel activities. These subsidies can hinder the transition to cleaner energy sources, as they make fossil fuels artificially cheap and artificially inflate the costs of renewable energy technologies.

Overall, it is evident that the fossil fuel industry has played a crucial role in driving global development and meeting growing energy demands. However, its significance is now being reevaluated in the face of escalating climate change. Actions to reduce reliance on fossil fuels, increase the share of renewable energy, and promote energy efficiency have become imperative to decarbonizing our energy systems and mitigating the impacts of climate change on a global scale.

Chapter 2: Origins of the Fossil Fuel Industry

The fossil fuel industry, comprising coal, oil, and natural gas, has played a crucial role in shaping our modern world. This chapter dives deep into the origins of this industry, exploring its origins, growth, and implications for human development and the environment.

1. Coal: An Ancient Energy Source

Coal, one of the oldest fossil fuels, has been used for centuries. In ancient China, coal was used for cooking and heating, while in 12th-century England, coal mining gained momentum. With the advent of the Industrial Revolution in the 18th century, the demand for coal skyrocketed. Its abundant availability and high energy density made it an ideal fuel for powering steam engines and supporting industrial processes.

2. The Birth of the Oil Industry

While coal dominated the early stages of the industrial era, the discovery of oil opened up new vistas for energy production. The first commercial oil well was drilled in Titusville, Pennsylvania in 1859, sparking a revolution in transportation and industrialization. Oil refining techniques improved rapidly, leading to the development of kerosene, a lighter fuel that replaced whale oil in lamps. Eventually, the combustion engine's invention in the late 19th century added another dimension to oil's importance, as it provided the primary energy source for automobiles.

3. Natural Gas Emerges as the Third Pillar

Although historically overshadowed by coal and oil, natural gas has played a vital role in the fossil fuel industry's development. As early as the ancient Greeks, natural gas seeping from the earth's surface was used for lighting and heating. However, its commercial exploitation required advancements in drilling and distribution technologies, which occurred during the mid-19th century.

4. Exploration and Extraction Technologies

The advancements in exploration and extraction technologies over time deserve special attention. Initially, coal was shallow-mined by hand, but with the Industrial Revolution, deeper and more extensive mining operations became the norm. The discovery of techniques for drilling oil wells deeper into the earth and on offshore platforms allowed access to oil and gas reservoirs. The introduction of fracking in recent decades unlocked previously inaccessible sources of natural gas trapped in shale formations.

5. Fossil Fuels and Human Development

The growth of the fossil fuel industry has undoubtedly driven human development. The abundant and relatively cheap energy sources provided by coal, oil, and gas have fueled industrialization, improved living standards, and supported economic growth worldwide. They have powered transportation, industries, and facilitated technological innovations. However, this rapid development has come with environmental and social consequences.

6. Environmental Impact

The burning of fossil fuels has released massive amounts of carbon dioxide $(CO2)$ into the atmosphere, exacerbating the greenhouse effect and contributing to climate change. Additionally, coal mining and oil extraction have scarred landscapes, damaged ecosystems, and threatened biodiversity. Spills, leaks, and accidents associated with the fossil fuel industry have led to water and soil pollution, devastating local communities and ecosystems.

7. Challenges and Transition Opportunities

The consequences of our heavy reliance on fossil fuels are increasingly apparent, leading to a global movement towards a transition to cleaner and more sustainable energy sources. Renewable energy technologies such as solar, wind, and hydropower have gained momentum in recent years. The phase-out of coal-fired power plants is becoming a priority in many countries, while efforts are being made to improve the efficiency and reduce the environmental impact of extracting and refining oil and gas.

THE FOSSIL FUEL INDUSTRY'S origins lie in the exploitation of coal, oil, and natural gas, which have been instrumental in driving human development while contributing to environmental challenges. Understanding the historical context, innovations, and impacts of this critical industry is crucial in shaping a sustainable energy future. By acknowledging the challenges and embracing transition opportunities, we can pave the way for a cleaner and more resilient energy system for generations to come.

- Evolution and development of the fossil fuel industry

The evolution and development of the fossil fuel industry have played a significant role in shaping the modern world. From its humble beginnings to its widespread global reach, this industry has had a profound impact on various aspects of human civilization. In this essay, we will delve into the rich history of the fossil fuel industry, exploring its origins, evolution, and the consequences of its development.

The story begins millions of years ago, during the geological processes that led to the formation of fossil fuels. Fossil fuels, such as coal, oil, and natural gas, are the remnants of ancient plants and organisms that were buried and subjected to immense pressure and heat underground over hundreds of millions of years. These resources, rich in carbon, have become essential sources of energy for human civilization.

Coal was the first fossil fuel to be widely used by humans. Its use dates back thousands of years, with historical records indicating its utilization for heating, cooking, and smelting purposes. However, it was during the Industrial Revolution in the late 18th and early 19th centuries that the coal industry experienced rapid growth. The invention of the steam engine by James Watt revolutionized transportation, manufacturing, and agriculture, providing an unprecedented demand for coal. This demand led to the emergence of coal mining as a distinct industry, with extensive networks of mines and railways established to extract and transport the resource.

The next major development in the fossil fuel industry came with the discovery and commercial exploitation of oil. In the mid-19th century, pioneers such as Edwin Drake drilled the first successful oil well in Pennsylvania, marking the beginning of a new era. Oil quickly proved to be a far more versatile and efficient source of energy compared to coal. It opened up new possibilities for transportation, providing fuel for automobiles, ships, and airplanes. The pharmaceutical and petrochemical industries also emerged, harnessing the vast potential of oil as a raw material.

BURNING EARTH: UNVEILING THE FLAMES

Natural gas, the third pillar of the fossil fuel industry, came into prominence in the 20th century. Although it had been used for centuries in small quantities, the widespread use of natural gas was made possible by advancements in extraction and transportation technologies. It became a valuable source of heating and electricity generation, with pipelines and liquefied natural gas (LNG) facilities established to increase its accessibility.

The 20th century witnessed a period of unprecedented growth for the fossil fuel industry. With the rise of industrialization and increasing global energy demands, fossil fuels became the dominant source of power worldwide. Major oil companies emerged, such as ExxonMobil, Shell, BP, and Chevron, exploiting resources on a massive scale. Cartels, like the Organization of the Petroleum Exporting Countries (OPEC), formed to regulate oil production and prices.

However, the rapid expansion and dependence on fossil fuels had unintended consequences. Environmental concerns began to surface, particularly with the discovery of the greenhouse effect and the potential for global climate change caused by carbon dioxide emissions. The negative impact on air quality through pollution and the depletion of finite resources further fueled these concerns.

In recent years, the fossil fuel industry has faced increasing scrutiny and challenges. Awareness of the environmental consequences of greenhouse gas emissions has led to efforts to transition to cleaner, renewable energy sources, such as solar, wind, and hydro. An international movement advocating for divestment from fossil fuels has gained momentum, pressuring companies and governments to take action.

In conclusion, the evolution and development of the fossil fuel industry have been marked by remarkable achievements and significant consequences. From coal to oil to natural gas, these resources have fueled economic growth, transformed societies, and shaped our modern world. However, their extensive use has also engendered environmental concerns that demand alternative approaches to energy generation. The path forward lies in balancing the need for energy with sustainable practices and embracing clean, renewable sources to power the future.

- Major players and key historical events that shaped the industry

The tech industry has seen significant growth and development over the years, with major players and key historical events that have shaped the industry into what it is today. From the rise of iconic companies to breakthrough innovations and game-changing acquisitions, the tech industry has constantly evolved and shaped our lives in profound ways.

One of the earliest major players in the tech industry was IBM (International Business Machines Corporation), which was founded in 1911. IBM played a significant role in computer manufacturing and hardware development, becoming a prominent player in the industry. They developed the IBM System/360, a groundbreaking mainframe computer series that revolutionized the way organizations processed data.

The 1980s witnessed the rise of Microsoft, founded by Bill Gates and Paul Allen. With the release of MS-DOS (Microsoft Disk Operating System), the company solidified its position as a key player in the personal computer industry. Their subsequent release of Windows, a user-friendly operating system, helped establish Microsoft as an industry leader and transformed the way people interacted with computers.

The 1990s brought about the internet boom, and several major players emerged. Among them was Netscape Communications Corporation, founded by Marc Andreessen and Jim Clark. Netscape Navigator, their flagship web browser, played a crucial role in popularizing the internet among mainstream users. This ultimately led to the dot-com bubble, where internet-based companies saw extraordinary valuations and investments, only to face a massive crash in the early 2000s.

Another significant player in the industry is Apple Inc., founded by Steve Jobs, Steve Wozniak, and Ronald Wayne in 1976. Initially gaining recognition for the Apple II, an early personal computer, Apple faced severe competition from PC manufacturers in the 1990s. However, the company witnessed a resurgence in the late 1990s with the introduction of the iMac, followed by groundbreaking innovations like the iPod, iPhone, and iPad, as well as the establishment of the Apple App Store.

Google, founded by Larry Page and Sergey Brin in 1998, completely revolutionized the search engine industry. Their algorithmic approach to search coupled with a simple user interface allowed users to find information quickly and easily. Moreover, Google expanded its portfolio, offering several innovative products and services like Gmail, Google Maps, and Google Drive. Google's dominance in online advertising and its Android operating system for mobile devices solidified its position as a major player in the tech industry.

In recent years, the industry has witnessed the rise of social media giants such as Facebook, founded by Mark Zuckerberg in 2004. Facebook has reshaped how we connect with friends, share information, and consume media. It has been instrumental in bringing social networking into the mainstream and has acquired other major players, like Instagram and WhatsApp.

In addition to major players, key historical events have also shaped the industry. The dot-com bubble burst in the early 2000s was a significant event that led to the collapse of many internet-based companies and a significant decline in investments. However, this event also led to the emergence of more sustainable business models that shaped the internet industry in the years to come.

Moreover, the introduction of smartphones and the subsequent app revolution transformed how we use technology in our daily lives. Apple's iPhone and the development of the App Store paved the way for a new era of mobile applications, allowing developers to innovate and create an entirely new ecosystem of services and experiences.

The industry is also notorious for notable acquisitions that have shaped its landscape. One such acquisition was Microsoft's purchase of LinkedIn, a professional networking platform, in 2016. This acquisition increased Microsoft's reach into social media and provided significant synergies between their existing products and LinkedIn's user base.

Overall, the tech industry has been shaped by major players like IBM, Microsoft, Apple, Google, and Facebook, and significant historical events such as the dot-com bubble and the rise of smartphones. These players and events continue to influence the industry and drive further innovation and development in key areas such as hardware, software, and internet services.

Chapter 3: Historical Debate on Global Warming

The global issue of climate change has long been a subject of heated debate and controversy. While there is a general consensus among the scientific community that the Earth's climate is indeed changing and that human activity is a significant contributing factor, this was not always the case. Chapter 3 delves deep into the historical debate surrounding global warming, providing readers with detailed and intriguing information that sheds light on the diverse perspectives that have shaped this discourse over time.

Point 1: The Early Skeptics

The historical debate on global warming can be traced back to the 19th century when scientists first began to notice potential shifts in climate patterns. However, it was not until the mid-20th century that the skeptic voices began to gain prominence. Scientists such as Frederick Seitz and Richard Lindzen expressed doubts regarding the impact of human activity on global temperatures, suggesting that natural variability could be the primary driver of climate change. Chapter 3 offers a comprehensive overview of the arguments put forth by these early skeptics and delves into the historical context that influenced their views.

Point 2: The Emergence of Consensus

Despite the skepticism that emerged in the 20th century, the majority of scientists gradually reached a consensus regarding the reality of human-induced global warming. Chapter 3 provides an in-depth analysis of the factors that led to this shift in scientific opinion. The emergence of advanced computer models, improved data collection techniques, and an extensive body of documented evidence all played a crucial role in laying the groundwork for a consensus among scientists. Moreover, this section examines the role of influential scientific reports, such as the Intergovernmental Panel on Climate Change (IPCC) assessments, in shaping the global understanding of climate change.

Point 3: Politicization and Debates

As global warming gained public attention, it quickly became a highly politicized issue, polarizing opinions across the political spectrum. Chapter 3 explores the diverse stakeholders involved in the political debate on climate change, diving into the complexity of their motivations and tactical strategies. This includes an examination of political think tanks, interest groups, and the usage of media to advance particular agendas. Additionally, this section covers significant events, such as the controversial "Climate-gate" scandal, which contributed to the intensification of the climate change debate.

Point 4: Implications for Public Policy

The historical debate on global warming profoundly impacted public policy decisions made by governments worldwide. Chapter 3 delves into the intricate relationship between science, politics, and policy-making. This section critically assesses the challenges faced by policymakers when dealing with such a complex and contentious issue. It explores the opposing perspectives on the urgency of response measures, the economic considerations involved, and the evolution of international efforts like the United Nations Framework Convention on Climate Change (UNFCCC) and the Paris Agreement.

CHAPTER 3 PROVIDES readers with a comprehensive and illuminating account of the historical debate surrounding global warming. By delving into the skeptical voices of the past, the emergence of consensus among the scientific community, the political polarization, and the implications for public policy, readers gain a rich understanding of the diverse perspectives that have shaped this crucial global issue. This chapter sets the stage for the subsequent exploration of solutions and actions undertaken to address climate change.

- Examination of early scientific discussions about global warming

During the past few decades, there has been growing concern about the impact of human activities on the Earth's climate. This concern has led to intense debates and discussions about global warming, its causes, and its potential effects. However, scientific discussions about global warming did not emerge overnight. They can be traced back to the early 19th century when scientists first started to explore the relationship between greenhouse gases and the Earth's temperature.

One of the earliest scientific discussions about global warming came from French mathematician Joseph Fourier in 1824. Fourier proposed that the Earth's atmosphere acts as a "blanket," trapping heat from the sun and preventing it from escaping into space. He suggested that changes in the composition of the atmosphere could affect this heat-trapping ability and subsequently alter the Earth's temperature.

Fourier's work laid the foundation for further study in this area. In the latter half of the 19th century, Irish physicist John Tyndall made significant contributions to the understanding of global warming. Tyndall conducted experiments to measure the heat-absorbing ability of different gases, including carbon dioxide (CO_2) and water vapor. He discovered that these gases have the ability to absorb infrared radiation, which is responsible for keeping the Earth warm.

Tyndall's experiments provided empirical evidence for the greenhouse effect and suggested that increasing concentrations of greenhouse gases could lead to a rise in global temperatures. His work was widely recognized, and it prompted further scientific inquiries into the sources and effects of possible human-induced climate change.

The late 19th and early 20th centuries saw continued progress in scientific discussions about global warming. Swedish chemist Svante Arrhenius published a seminal paper in 1896 in which he calculated the potential increases in global temperature resulting from increased levels of CO_2 in the atmosphere. Although Arrhenius had limited data to work with, he predicted that a doubling of atmospheric CO2 levels would lead to a global temperature increase of about 4-6 degrees Celsius.

Arrhenius's calculations were significant not only because they provided early estimates of the magnitude of the greenhouse effect but also because they highlighted the potential human contribution to global warming. He suggested that the burning of fossil fuels, such as coal and oil, could lead to an increase in atmospheric CO2 levels, thereby causing global warming.

As the scientific community's understanding of global warming grew, so did the complexity of the discussions surrounding it. One key development in the mid-20th century was the recognition of the role of feedback mechanisms in climate change. Scientists recognized that changes in temperature could trigger various positive and negative feedback loops, further amplifying or dampening the effects of global warming.

For instance, rising temperatures could lead to increased evaporation, which, in turn, could result in higher concentrations of water vapor—a potent greenhouse gas. Similarly, the melting of ice caps could reduce the Earth's albedo, or reflectivity, thus exacerbating global warming.

With advancements in computer technology in the latter half of the 20th century, scientists began to develop sophisticated climate models. These models allowed scientists to simulate various climate scenarios, exploring the potential consequences of different greenhouse gas emissions and other drivers of climate change.

It is important to note that while scientific discussions about global warming have made significant progress over the centuries, there have also been challenges and disagreements along the way. Skepticism and debate about the causes and magnitude of global warming have been persistent within the scientific community.

Nevertheless, the examination of early scientific discussions about global warming provides crucial insights into how our understanding of this complex phenomenon has evolved over time. From Fourier's early recognition of the greenhouse effect to Arrhenius's calculation of the potential consequences of increasing CO2 levels, these pioneers laid the groundwork for the scientific consensus we have today—that human activities are driving global warming and climate change.

In conclusion, the examination of early scientific discussions about global warming reveals the long and detailed path scientists have taken to understand this complex issue. From Fourier to Tyndall to Arrhenius, each contribution has built upon earlier discoveries, gradually shaping our understanding of the causes, mechanisms, and potential impacts of global warming. The ongoing debates and disagreements serve as a reminder of the complexity and ongoing research in this field, further motivating us to take appropriate actions to mitigate the effects of climate change.

- Initial understanding and skepticism surrounding the concept

In the world of innovation and new ideas, there are always initial moments of understanding and skepticism. This is particularly true when it comes to revolutionary concepts that challenge existing norms and beliefs. Such is the case with the concept we are going to delve into in this piece.

The initial understanding of this concept may vary from person to person. Some might have an immediate grasp of its potential, while others may struggle to wrap their heads around it. The complexity and intricacy of the concept may contribute to this dichotomy in comprehension.

When a new concept is introduced, the first hurdle to overcome is grasping the basic premise. The initial description might seem complex with unfamiliar terms, making it a challenging task to achieve full understanding. But as with anything, patience and perseverance can lead to a breakthrough.

Some individuals may also approach the concept with skepticism right from the outset. This skepticism could stem from various factors. One reason could be a deeply-rooted attachment to existing beliefs and ideas, making it difficult to accept something completely new and different. Others may be skeptical due to a lack of knowledge or understanding, preferring to maintain a distance from what they perceive as a risky or confusing proposition.

Nevertheless, this skepticism is not necessarily a bad thing. In fact, it is a vital part of exploring new concepts and ideas. It is through skepticism that we ensure thorough exploration and evaluation, ultimately leading to a more solid conclusion.

As time progresses, and more information is shared and discussed, the understanding around the concept begins to solidify. Some individuals who were initially skeptical may begin to see the potential and validity of the idea. This transition often occurs through engaging in further research, conversations, and exposure to real-world applications.

Interestingly, the more detailed and intricate the information surrounding the concept becomes, the more captivating it becomes. The layers of complexity begin to unravel, revealing a fascinating tapestry of innovation and possibility. People naturally gravitate towards such depth, thirsting for the intricacies that underpin revolutionary ideas.

The evolution of understanding and skepticism ultimately contributes to the development and refinement of the concept. As individuals question and probe, new perspectives emerge, and the concept continues to grow and adapt. It is through this iterative process of questioning, understanding, and skepticism that groundbreaking ideas find their footing and lead to tangible advancements.

In conclusion, the journey from initial understanding and skepticism to embracing a new concept is a fascinating and pivotal one. The complexity and detail surrounding the concept, combined with the natural skepticism that arises, contribute to an enriching and sometimes challenging exploration. However, as understanding deepens and skepticism is explored further, the concept gains momentum and begins to transform from an abstract idea to a tangible reality.

Chapter 4: Impact of Fossil Fuels on Global Warming

Fossil fuels have been instrumental in powering the modern industrialized world. They have contributed significantly to economic development and improved quality of life. However, the extensive use of these non-renewable energy sources comes with severe environmental consequences, most notably global warming. In this chapter, we will delve into the intricate relationship between fossil fuels and the impact they have on global warming, exploring the underlying mechanisms and the consequences they present for the planet.

1. Greenhouse Effect:

To understand the impact of fossil fuels on global warming, it is essential to grasp the concept of the greenhouse effect. The Earth's atmosphere acts as a protective shield, allowing sunlight to enter but trapping a portion of the heat within. This is beneficial, as it sustains habitable conditions on our planet. However, the excessive burning of fossil fuels leads to excessive release of greenhouse gases, such as carbon dioxide (CO_2), methane (CH_4), and nitrous oxide (N_2O), which enhance the natural greenhouse effect. Over time, the increased concentration of these gases results in a gradual rise in Earth's temperature, leading to global warming.

2. CO2 Emissions from Fossil Fuels:

Carbon dioxide is the primary greenhouse gas emitted during fossil fuel combustion. Burning fossil fuels, such as coal, oil, and natural gas, releases vast amounts of carbon dioxide into the atmosphere. The combustion of coal releases the highest amount of CO_2 per unit of energy produced, followed by oil and then natural gas. These emissions primarily result from energy production, industrial processes, and transportation.

3. Methane and Nitrous Oxide Emissions:

Although carbon dioxide is the most abundant greenhouse gas emitted by fossil fuels, methane and nitrous oxide also contribute to global warming. Methane is released during the extraction, transportation, and use of fossil fuels, especially from leaking natural gas pipelines. Additionally, methane is produced by livestock and other agricultural activities, including the decomposition of organic waste. Nitrous oxide, on the other hand, originates from agricultural and industrial activities, mainly due to the use of nitrogen-based fertilizers and the combustion of fossil fuels.

4. Deforestation and Fossil Fuels:

The relationship between fossil fuels and global warming extends beyond emissions. Deforestation, which occurs predominantly to clear land for agriculture and the production of commodities like palm oil and soy, significantly exacerbates the impact of greenhouse gases. Trees play a crucial role in absorbing carbon dioxide from the atmosphere through photosynthesis, thus acting as natural carbon sinks. However, deforestation reduces the planet's ability to absorb CO_2, releasing stored carbon and contributing to the escalating levels of this gas.

5. Consequences of Global Warming:

The impact of fossil fuels on global warming leads to a range of adverse consequences. Rising average temperatures disrupt ecosystems and natural weather patterns, resulting in extreme weather events like hurricanes, heatwaves, and droughts. Sea levels are also rising due to the melting of polar ice caps, endangering coastal communities and low-lying regions. Furthermore, global warming is affecting agricultural productivity, leading to food security concerns as changes in temperature and precipitation patterns impact crop yields.

THE DETAILED EXPLORATION of the impact of fossil fuels on global warming reveals a complex and intertwined relationship. The extensive use of fossil fuels, coupled with deforestation, releases significant amounts of greenhouse gases into the atmosphere, intensifying the greenhouse effect and driving climate change. It is becoming increasingly critical to employ sustainable and greener alternatives to fossil fuels, ensuring a healthy and stable environment for future generations. Our awareness and action are paramount in mitigating the adverse effects of fossil fuel-induced global warming.

- Scientific evidence linking the burning of fossil fuels to climate change

Scientific evidence linking the burning of fossil fuels to climate change is a complex and crucial topic that has garnered substantial attention in recent years. The understanding of this connection relies on a multitude of data sets, studies, and observations made by scientists around the globe. In this article, we will explore the overwhelming evidence that supports this link, highlighting key studies and their findings.

One of the most significant pieces of evidence is taking stock of greenhouse gas concentrations in the atmosphere. The burning of fossil fuels releases vast amounts of carbon dioxide (CO_2), a potent greenhouse gas that traps heat in the atmosphere, leading to the phenomenon known as the greenhouse effect. Since the Industrial Revolution, there has been a significant increase in atmospheric CO_2 concentrations, primarily driven by human activities like the burning of coal, oil, and natural gas. This increase can be traced back to industrial and technological advancements that rely heavily on fossil fuels.

The Keeling Curve, a widely referenced graph, shows year-on-year increases in atmospheric CO_2 levels observed at Mauna Loa, Hawaii. Initiated by the late scientist Charles David Keeling in 1958, this ongoing monitoring effort has consistently demonstrated an upwards trend in CO_2 concentrations. Currently, CO_2 levels have exceeded 410 parts per million (ppm), far above pre-industrial levels of around 280 ppm. This unprecedented rise correlates closely with the increased combustion of fossil fuels during this period.

However, mere rises in CO_2 concentrations alone do not directly prove a link between fossil fuel burning and climate change. To establish causation, scientists have conducted numerous studies exploring the relationship between these emissions and changes in global temperature. One prominent example is the Intergovernmental Panel on Climate Change (IPCC) assessment reports, which compile the work of thousands of scientists worldwide. Their fifth assessment report in 2014 concluded with high confidence that human activities, particularly the burning of fossil fuels, are the dominant cause of the observed warming since the mid-20th century.

In this report, the IPCC examined various lines of evidence, including sophisticated climate models, historical temperature records, and natural climate variability. Through these analyses, scientists were able to attribute the significant temperature increases in recent decades to anthropogenic factors. The IPCC report also highlighted the fact that natural factors such as solar radiation and volcanic eruptions alone cannot explain the observed warming trend.

Additionally, instrumental records provide essential evidence in linking fossil fuel burning to climate change. By examining temperature records, ice core data, and other measurements, scientists have unequivocally shown a close correlation between rising greenhouse gas emissions and global temperature increases over the last century. This correlation strengthens the case that anthropogenic activities, particularly fossil fuel emissions, are driving climate change.

Moreover, researchers have employed a range of indirect methods to reconstruct past climate conditions, such as studying ice cores, tree rings, and sedimentary deposits. These proxies can provide insight into both natural climate cycles and how humans have influenced the climate system. Analyses of these indicators consistently point to the burning of fossil fuels as a significant driver of modern climate change, with unprecedented changes observed in recent centuries.

In summary, the scientific evidence linking the burning of fossil fuels to climate change is abundantly clear and supported by a diverse array of research. Rising carbon dioxide levels, backed by the Keeling Curve and other measurements, serve as distinct markers of human activities. The IPCC synthesis reports offer an extensive body of scientific knowledge demonstrating the robust connection between anthropogenic greenhouse gas emissions and observed global warming. Moreover, instrumental records and proxy data consistently illustrate the strong correlation between increased fossil fuel combustion and alterations in the Earth's climate system. Collectively, these findings underline the urgent need to transition to cleaner energy sources and mitigate the harmful impacts of climate change.

- Overview of greenhouse gas emissions and their consequences

Greenhouse gas emissions refer to the release of gases into the atmosphere that trap heat and contribute to the greenhouse effect, leading to global warming and climate change. These gases include carbon dioxide (CO_2), methane (CH_4), nitrous oxides (N_2O), and fluorinated gases. While some greenhouse gas emissions occur naturally, human activities have significantly increased their levels in the atmosphere, primarily through the burning of fossil fuels, deforestation, and industrial processes.

The consequences of greenhouse gas emissions are far-reaching and impact various aspects of life on Earth. The most significant consequence is global warming, which involves an increase in the average temperature of the Earth's surface and oceans. This warming trend leads to a range of other environmental changes and challenges.

One prominent consequence is the rise in sea levels as a result of melting ice caps and thermal expansion of seawater. Rising sea levels threaten coastal cities and low-lying regions, leading to increased flooding, erosion, and loss of habitat for plants and animals. Small island nations particularly face the risk of total submersion.

Additionally, changing weather patterns and increased frequency and severity of extreme weather events are consequences of greenhouse gas emissions. This includes heatwaves, droughts, hurricanes, and heavy rainfall events. These changes in weather patterns can have devastating effects on ecosystems, agriculture, water resources, infrastructure, and human health.

The consequences of greenhouse gas emissions are not confined to specific regions but have a global impact. Furthermore, they exacerbate existing environmental problems and contribute to biodiversity loss, as many species struggle to adapt quickly enough to changing conditions.

In addition to the environmental effects, greenhouse gas emissions also have economic and social consequences. These consequences range from increased healthcare costs due to air pollution and heat-related illnesses to reduced agricultural productivity and food security. Communities dependent on natural resources, such as farmers, fishermen, and indigenous populations, are particularly vulnerable to the impacts of climate change.

Addressing greenhouse gas emissions is paramount to mitigate the consequences and maintain a sustainable future. This includes reducing our reliance on fossil fuels through transitioning to renewable energy sources, improving energy efficiency, promoting sustainable land management practices, and adopting greener technologies. International cooperation and global efforts are essential to combat the causes of greenhouse gas emissions and ensure a stable climate for future generations. Numerous international agreements, such as the Paris Agreement, aim to limit global warming below 2 degrees Celsius above pre-industrial levels.

In conclusion, greenhouse gas emissions have severe consequences that affect various aspects of life on Earth, including the environment, economy, and society. Understanding and addressing these emissions is crucial to combat climate change, protect vulnerable communities and ecosystems, and ensure a sustainable future for our planet.

Chapter 5: Fossil Fuel Industry's Response to Global Warming Findings

As awareness and understanding of the impact of human activities on global warming increased, so did the need for industries heavily reliant on fossil fuels, namely the fossil fuel industry, to acknowledge and address these findings. This chapter delves into the response of the fossil fuel industry to global warming findings, examining its early denial and subsequent actions to counteract the growing evidence of their contributions to climate change.

Early Denial:

In the initial stages of scientific discussions surrounding climate change, the fossil fuel industry often adopted a defensive stance, questioning the validity of research linking their activities to global warming. Industry-funded studies denying the human impact on the climate were conducted, leading to public confusion and delay in the adoption of necessary policies to mitigate climate change. Several prominent individuals and organizations within the industry actively supported and propagated these contradictory arguments, casting doubt on the urgency of addressing global warming.

Shifting Narrative:

As scientific consensus solidified around the human influence on climate change, the fossil fuel industry began recognizing the need for a change in strategy. They shifted their focus towards challenging the extent of their contribution to global warming, emphasizing individual responsibility rather than systemic change. Ad campaigns, sponsored research, and lobbying efforts aimed to undermine policy initiatives that sought to regulate emissions, emphasizing the importance of diverse energy sources to global development.

Public Relations Efforts:

Aware of the potential reputational risks associated with being labeled as climate villains, fossil fuel companies invested heavily in public relations campaigns. These endeavors promoted the companies as responsible environmental stewards engaged in research and innovation aimed at delivering sustainable solutions. Sponsorships of environmental projects, public forums, and scientific events were used to showcase commitment to finding alternative fuels as well as downplaying the impact of their core business practices on global warming.

Co-opting Climate Science:

Recognizing the increasing societal and political pressure to move away from fossil fuels, the industry pursued collaborative approaches with scientific institutions, aiming to modify the public's perspective about their practices. By funding research centers and institutes specializing in climate studies, fossil fuel companies gained leverage in influencing the scientific discourse. This strategy allowed them to partially regain public trust but also brought concerns over compromised integrity and potential conflicts of interest.

Delaying Climate Policy:

Behind the scenes, the fossil fuel industry actively engaged in lobbying efforts to delay or dilute environmental regulations aimed at curbing greenhouse gas emissions. By employing an array of tactics, including the promotion of uncertainty surrounding climate science, financial support to climate-skeptic politicians, and leveraging their economic influence to sway policymakers, the industry has successfully hindered progress on climate policy for decades.

THE FOSSIL FUEL INDUSTRY'S response to global warming findings demonstrates a complex and multifaceted approach aimed at preserving its interests and perpetuating the dominance of fossil fuels in the energy sector. From initial denial to narrative shifting, public relations efforts, co-opting climate science, and delaying policy, the industry has shown an ability to adapt to changing circumstances while maintaining its influential position. Addressing climate change requires recognizing the role played by such industries and fostering transparent and proactive engagement to ensure a sustainable future.

- Analysis of the industry's initial conduct and reaction to scientific insights

The initial conduct and reaction of industries to scientific insights can reveal a lot about their interests, intentions, and commitment to societal well-being. As scientific breakthroughs occur, it is crucial for industries to promptly recognize and respond to these insights in a responsible and ethical manner.

One aspect of analyzing the industry's initial conduct is examining how they embrace or resist scientific discoveries. Industries that prioritize growth and innovation are more likely to recognize the value of scientific insights and incorporate them into their operations. For instance, industries in the technology sector have consistently embraced scientific advancements, utilizing them to create cutting-edge products and services that cater to evolving consumer demands. These industries leverage scientific research to enhance their competitiveness, staying ahead of the curve.

Conversely, industries that rely heavily on outdated practices or exhibit resistance to change often struggle in incorporating new scientific insights. Energy sectors, for example, have faced varying degrees of resistance towards renewable energy sources due to their traditional reliance on fossil fuels. Despite renewable energy's potential for environment-friendly operations, some key players in these industries have been slow to transition and instead focused on preserving their current business models. Hence, investigating industry-specific responses can shed light on the motivations underlying their initial conduct.

Moreover, analyzing how the industry communicates and disseminates scientific findings is critical. Industries that prioritize transparency and open communication are more likely to act responsibly in light of new scientific insights. They actively engage with the scientific community, collaborate in research endeavors, and encourage scientific literacy among their employees and stakeholders.

On the other hand, industries that lack transparency may downplay or manipulate scientific research to safeguard their financial interests. Such conduct can undermine public trust and hinder the adoption and application of scientific discoveries. A classic example is the tobacco industry, which, despite years of scientific evidence linking smoking to severe health risks, took various measures to mislead the public and avoid regulation for as long as possible.

Furthermore, the industry's initial response to scientific insights can also reveal their commitment to ethical practices and corporate social responsibility. Responsible industries proactively prioritize the health and safety of consumers and the environment, incorporating scientific insights into their production processes, supply chains, and product development. Pharmaceutical and food industries, for instance, heavily rely on scientific research to ensure compliance with regulatory standards and deliver safe and effective products to consumers.

Conversely, the industries that exhibit a negligent or reactive approach to incorporating scientific insights may compromise the well-being and trust of their constituents. This negligence can manifest in industries ignoring or failing to address scientific evidence related to negative health effects, environmental damages, or workplace safety concerns, among others. Such conduct reveals a lack of commitment to the greater good and may attract public backlash, legal liabilities, and long-term reputational damage.

Overall, analyzing the industry's initial conduct and reaction to scientific insights provides valuable insights into their priorities, strategies and adherence to ethical practices. Industries that enthusiastically embrace scientific breakthroughs, prioritize transparency, and demonstrate a commitment to societal well-being, are more likely to foster innovation and sustainable growth while contributing positively to society. Understanding the conduct and reaction of industries towards scientific insights can help policymakers, consumers, and stakeholders in evaluating the level of responsibility and integrity they exhibit in the face of ongoing scientific advancements.

- Influence of groups within the industry and climate change denial

I nfluence of groups within the industry and climate change denial
The issue of climate change has long been a topic of concern for scientists, politicians, and individuals around the world. As the scientific consensus on the existence and impact of climate change has grown, so has the response from various industries and interest groups. While many organizations have taken steps to mitigate their environmental impact and support efforts to combat climate change, others have actively sowed doubt and denial regarding the reality and severity of the issue. This article aims to explore the influence of these groups within the industry and their role in climate change denial.

One of the most prominent groups involved in climate change denial is the fossil fuel industry. As the primary beneficiaries of the global reliance on fossil fuels, these companies have a strong vested interest in downplaying the issue of climate change and perpetuating the status quo. Through their extensive lobbying efforts, campaign donations, and public relations campaigns, these groups have successfully shaped public perception and influenced policy decisions. They have funded and promoted research studies casting doubt on the scientific consensus, employed tactics to discredit climate scientists, and funded think tanks to produce climate change skeptics.

In addition to the fossil fuel industry, conservative think tanks and advocacy groups have played a significant role in climate change denial. These organizations promote free-market policies and limited government intervention, often aligning with the interests of the fossil fuel industry. By framing climate change as a threat to economic growth and individual freedom, these groups have been successful in shaping public opinion and expanding doubt about climate change. They have positioned themselves as alternative sources of scientific information, disputing the widely accepted consensus reached by the scientific community.

Media organizations, particularly those with conservative political leanings, have also played a pivotal role in amplifying climate change denial. Fox News, for instance, has consistently provided a platform for climate change skeptics, subtly sowing doubt by presenting opposing viewpoints without proper context and scientific scrutiny. Additionally, social media platforms have been instrumental in the dissemination of misinformation and conspiracy theories related to climate change. Fake news sites, echo chambers, and the algorithms underlying these platforms have resulted in the creation of filter bubbles, where users are exposed to information that confirms their existing beliefs, reinforcing climate change denialism.

The influence of these groups within the industry has extended far beyond shaping public opinion. Their efforts have profoundly affected climate change policy decisions at both national and international levels. By systematically challenging climate science, they have effectively delayed and weakened policy implementation, hindering progress in mitigating greenhouse gas emissions and transitioning towards renewable energy sources. Their power is most evident in the United States, where climate change denial has been a prevailing issue, resulting in the country's withdrawal from international climate agreements and the overturning of domestic environmental regulations.

While the efforts of groups denying climate change have had significant impact, it is important to recognize the growing counter-movement working towards environmental sustainability. As the consequences of climate change become increasingly apparent and public concern grows, businesses, investors, and individuals are demanding more action and becoming advocates for change. Many organizations within the industry, including major fossil fuel companies, have begun to embrace renewable energy sources and acknowledge the need for addressing climate change.

In conclusion, the influence of groups within the industry on climate change denial cannot be underestimated. The fossil fuel industry, conservative think tanks, and media organizations have all contributed to shaping public perception and influencing policy decisions by sowing doubt and disputing the scientific consensus. Their efforts have successfully delayed progress in tackling climate change, hindering mitigation measures and impeding the transition to renewable energy sources. However, it is crucial to acknowledge the growing counter-movement and the increasing recognition of climate change as a critically important issue. Only by understanding the influence of these groups can we work towards meaningful change and create a sustainable future.

31

Chapter 6: Lobbying and Political Influence

Lobbying and political influence have long been a subject of interest and concern in democratic societies. Lobbying refers to activities aimed at influencing governmental decisions, policies, and legislation. It allows individuals, corporations, interest groups, and professional associations to present their perspectives and push for their preferred outcomes. This chapter aims to delve into the world of lobbying, exploring its practices, impacts, and controversies associated with political influence in democratic societies.

1. Understanding Lobbying:

1.1 Definition and Scope:

Lobbying, at its core, represents attempts to shape or influence public policy. It encompasses various activities, such as advocacy, information dissemination, research, public education, and political mobilization. These activities allow lobbyists to present their viewpoint directly to policymakers and urge them to consider their interests during decision-making processes.

1.2 Lobbyists and Lobbying Organizations:

Lobbyists can work independently or as part of lobbying organizations. They employ various tactics to advocate for their clients, such as meeting policymakers, providing expert advice, drafting legislation, organizing grassroots campaigns, hosting fundraisers, and engaging in public relations campaigns. Lobbying organizations, on the other hand, can range from trade associations and corporate interest groups to non-profit organizations and unions.

2. Lobbying Process:

2.1 Engaging Policymakers:

Successful lobbying necessitates building relationships with key policymakers who have decision-making authority. Lobbyists seek meetings, offer campaign contributions, and make use of social events to connect with politicians and bureaucrats. These interactions help establish trust, foster understanding, and ensure lobbyists have access to decision-making processes.

2.2 Influencing Public Opinion:

Public opinion plays a crucial role in shaping policy decisions. Lobbyists often employ advocacy, public relations, and media outreach strategies to gain public support for their cause. Through social media campaigns, press releases, and op-eds, lobbyists can shape the narrative surrounding their issues and generate public pressure on policymakers.

2.3 Preparing Policy Proposals:

Lobbyists often engage in thorough research and analysis to prepare detailed policy proposals. They may consult subject matter experts, economists, lawyers, and other professionals to develop proposals that align with their goals and satisfy the demands of policymakers. These proposals can directly influence the content and direction of legislation.

3. Controversies and Challenges:

3.1 Ethical Concerns:

Lobbying raises ethical questions regarding the influence of money in politics, the potential for corruption, and biased decision-making. Critics argue that lobbying can skew the democratic process by disproportionately favoring well-funded or powerful interest groups over the general public's interests. Transparency and disclosure requirements are therefore necessary to address these ethical concerns.

3.2 Regulatory Frameworks:

Governments strive to strike a balance between allowing lobbying as a democratic practice and minimizing potential abuses. Many countries have established laws and regulations with reporting requirements, spending limits, and disclosure rules. The creation of lobbying registries and mechanisms for public scrutiny helps enhance transparency, but challenges persist in effectively regulating and enforcing these rules.

3.3 Influence of Big Money:

The disproportionate influence of wealthy individuals and corporations has been a significant concern in lobbying. Critics argue that big money can unduly shape legislation and undermine democracy by affording direct access to decision-makers or indirectly influencing public discourse through funding media campaigns. Policymakers must remain diligent in guarding against any undue influence exerted by powerful interests.

LOBBYING AND POLITICAL influence continue to play a significant role in contemporary democratic societies. While lobbying mechanisms serve as channels for citizens and organizations to voice their concerns, they give rise to debates about ethics, transparency, and the appropriate regulation of political influence. A nuanced understanding of the lobbying process, its benefits, and its challenges is crucial for policymakers, citizens, and stakeholders to ensure democratic values are preserved and democratic decision-making remains robust and accountable.

- Exploration of the fossil fuel industry's lobbying efforts and political sway

The fossil fuel industry is known for its extensive lobbying efforts and political sway. Through these tactics, these companies wield considerable influence on policy-making, which in turn shapes the energy landscape and curtails the growth of renewable energy alternatives. This article will delve into the various aspects of the industry's lobbying tactics, its political allies, and the implications of its overall power.

First and foremost, lobbying is the primary mechanism through which the fossil fuel industry asserts its influence over government decision-making. Lobbying entails the act of advocating for particular policies, often through financial contributions to candidates, influencing public opinion, and leveraging close relationships with lawmakers. Fossil fuel companies invest substantial financial resources in lobbying, allowing them to access and shape policy discussions that directly impact their profits and operations.

One key tactic employed by the industry is campaign financing. Fossil fuel companies, including oil, gas, and coal firms, frequently donate large sums of money to political campaigns, particularly to candidates who champion their interests. Such contributions can sway elections in favor of the industry's allies, who, once elected, are more inclined to advocate for policies that further its agenda.

In addition to providing direct campaign financing, the fossil fuel industry also utilizes various indirect channels to influence public opinion and policy debates. One common strategy is funneling money to organizations and think tanks that promote climate change denial or work to undermine the scientific consensus on the causes and impacts of global warming. These groups use their platforms to disseminate misleading information, often casting doubt upon the urgency of transitioning away from fossil fuels.

Moreover, the industry enjoys close ties with key players in political circles. Many politicians and policymakers have past or ongoing connections to fossil fuel companies, either through running or working for these corporations, receiving campaign contributions, or collaborating on legislation. These relationships create a revolving door phenomenon, wherein individuals move effortlessly between the industry and government roles, serving to perpetuate the influence of fossil fuel interests on policy-making processes.

The consequences of the fossil fuel industry's lobbying efforts and political sway are far-reaching. Firstly, they impede the development and adoption of renewable energy technologies. By ensuring policies that favor fossil fuels, the industry consistently stifles the growth of clean alternatives, such as wind and solar power. This perpetuates our reliance on finite and environmentally damaging resources, hindering progress towards a more sustainable future.

Additionally, the industry's influence extends to climate change policy. Despite scientific consensus on the need to reduce greenhouse gas emissions, fossil fuel companies successfully lobbied to undermine international climate agreements, such as the Paris Agreement. Their tactics have included sowing doubt about the severity of climate change, lobbying against regulation, and investing in massive public relations campaigns to present themselves as environmentally responsible stewards.

Critics argue that the current fossil fuel-dominated energy system results from a rigged political playing field. They contend that harnessing the true potential of renewable energy sources requires not only technological advancements but also addressing the imbalance of power created by the fossil fuel industry's lobbyism and political clout.

In conclusion, the fossil fuel industry's lobbying efforts and political sway remain formidable obstacles to progress in combating climate change and advancing renewable energy. Through campaign contributions, indirect funding, and close connections with lawmakers, this industry shapes policy, suppresses alternatives, and perpetuates a reliance on environmentally damaging energy sources. Overcoming the exercise of the fossil fuel industry's power demands a concerted effort to break free from their influence and refocus political and legislative agendas towards sustainable, renewable energy solutions.

- Examination of policy decisions influenced by industry interests

E xamination of Policy Decisions Influenced by Industry Interests

POLICY DECISIONS PLAY a critical role in shaping societal outcomes, striking a delicate balance between various stakeholders' interests and welfare. However, the influence of industry interests on policy decisions has been a subject of scrutiny and discourse. This write-up aims to delve into the complexities and consequences associated with policies influenced by industry priorities. Through an exploration of various examples, we will examine the potential ramifications on public welfare and democratic processes.

Body:

I. Unintended Bias:

When industry interests heavily steer policy decisions, concerns regarding the lack of objectivity and impartiality often surface. Such influence may lead to policies crafted to favor particular sectors, causing unintended biases that undermine the overall public interest. For instance, pharmaceutical industry influence exerted on healthcare policy has raised concerns regarding inflated drug prices and hindered access to affordable medications.

II. Regulatory Capture:

The phenomenon of regulatory capture occurs when industry players shape and manipulate regulatory bodies and agencies meant to regulate their activities. This undue influence can limit the efficacy of regulations or delimit their scope, compromising the intended social and environmental goals. Examples of regulatory capture encompass industry-funded studies discrediting scientific evidence related to climate change and the influential lobbying by online short-term rental platforms, hindering local authorities' regulation efforts.

III. Suppression of Competitors:

Policy decisions influenced by industry interests may contribute to stifling competition and constraining innovation. By leveraging their sway, incumbent industry players may advocate for regulations that impede competitors' entry or growth, stifling market dynamics. This can impede economic progress and consumer choice, limiting a fair and flourishing marketplace of ideas.

IV. Dampening Public Health and Safety Efforts:

The interplay between industry interests and policy decisions related to public health and safety can significantly impact society. Industries with addictive products such as tobacco and sugary beverages have historically employed lobbying and influence to counter measures that aim to reduce health risks. This influence can perpetuate harmful practices, creating substantial long-term public health and safety concerns.

V. Democratic Processes:

The democratic processes governing policy formulation and decisions should encapsulate the diverse voices and concerns of the citizenry. However, when industry interests disproportionately influence policies, the democratic process is at risk of being undermined. The power of lobbying, campaign contributions, and revolving-door phenomena between industry and government can sway decision-making, potentially undermining the democratic ideal of governance for the people.

WHILE IT IS ESSENTIAL for policymakers to engage with industries and consider their perspectives, excessive influence can lead to policy decisions that prioritize narrow interests over the common good. This examination emphasizes the unintended biases, regulatory capture, suppression of competitors, dampening of public health efforts, and potential erosion of democratic processes accompanying policies influenced by industry interests. Acknowledging, discussing, and addressing these concerns is vital to fostering fair and equitable policy decisions that encompass the best interests of society as a whole.

Chapter 7: Economic Implications of Global Warming

Global warming, a phenomenon caused by the accumulation of greenhouse gases in the Earth's atmosphere, has far-reaching economic implications. The consequences of rising temperatures, changing climate patterns, and extreme weather events extend well beyond environmental concerns. In this chapter, we will explore the various economic impacts of global warming, spanning industries, regions, and economies worldwide.

Impact on Agriculture

One of the most significant economic implications of global warming is its impact on agriculture. Changes in rainfall patterns, rising temperatures, and increased frequency of extreme weather events present substantial challenges to crop production and livestock farming. Crop yields may decline, affecting food security, especially in vulnerable regions. Furthermore, the economic costs of adapting agricultural practices to the changing climate can be substantial, including investments in irrigation systems, infrastructure, and crop diversification.

Rising Sea Levels and Coastal Infrastructure

Global warming-induced rising sea levels pose a significant threat to coastal infrastructure. With an estimated one-third of the global population residing within 100 kilometers of the coast, the potential economic damage is immense. Submergence of coastal cities, erosion of coastlines, and increased instances of flooding can result in the destruction of infrastructure, relocation costs, and loss of livelihoods. Developing countries, without adequate resources for climate change adaptation, will be disproportionately impacted.

Energy and Fossil Fuels

Global warming necessitates a rapid transition to cleaner energy sources, making it imperative to reduce reliance on fossil fuels. Energy-intensive industries such as manufacturing and transportation face both challenges and opportunities in adapting to a low-carbon future. The economic implications of this transition involve high upfront costs for renewable energy infrastructure and the need for substantial public and private investments in research and development to drive innovation in clean technology.

International Trade and Supply Chains

Global warming-induced disruptions to international trade and supply chains can have significant economic consequences. Extreme weather events, such as hurricanes and floods, can damage critical infrastructure, leading to disruptions in the production and distribution of goods and services. This disruption cascades through supply chains, negatively impacting industries beyond those directly affected. As a result, the global economy can experience reduced efficiency and increased costs.

Tourism and Coastal Economies

The tourism industry, particularly coastal tourism, is highly vulnerable to global warming impacts. Rising temperatures can lead to the degradation of delicate ecosystems like coral reefs and loss of biodiversity, reducing the appeal of tourist destinations. Additionally, increased instances of extreme weather events can damage infrastructure and discourage travel. Coastal economies heavily reliant on tourism revenue can suffer significant economic setbacks if appropriate adaptation and mitigation measures are not taken.

Healthcare Expenditure

Global warming also has substantial implications for healthcare expenditure. Changing climate patterns, heatwaves, and increased prevalence of infectious diseases pose direct health risks to populations. Healthcare systems must redirect significant resources towards combating climate-related health issues, thereby reducing their capacity to fund other necessary healthcare services and initiatives. This diversion of funds can strain economies' overall spending and impede development efforts.

THE ECONOMIC IMPLICATIONS of global warming are vast and multifaceted, impacting diverse sectors, regions, and economies worldwide. While some industries may face challenges and high adaptation costs, there are also opportunities for innovation and economic growth through mitigation and cleaner technology adoption. Nonetheless, addressing global warming requires concerted efforts from governments, businesses, and individuals to reduce greenhouse gas emissions, adapt to changing climates, and build resilient economies for a sustainable future.

41

- Discussion on the economic consequences of climate change caused by fossil fuels

Climate change caused by the burning of fossil fuels has become one of the most significant challenges of our time. The consequences of this global crisis have the potential to affect various aspects of our lives and have severe economic implications. In this discussion, we will explore the economic consequences of climate change caused by fossil fuels and delve into the intricate details surrounding this pertinent issue.

One of the primary economic consequences of climate change is the increased frequency and intensity of extreme weather events. These events, such as hurricanes, floods, and droughts, have the potential to cause widespread destruction, resulting in substantial economic losses. Not only do these disasters lead to immediate costs in terms of property damage and human casualties, but they also have long-term effects on the affected regions.

This devastation can result in billions or even trillions of dollars spent on rebuilding efforts, often draining valuable resources from governments and diverting funds that could have been allocated to other social and economic development projects. Similarly, the impact on infrastructure, particularly in coastal areas, can result in significant expenses as coastal protections and rehabilitation becomes necessary. Thus, governments and societies are forced to divert funds into recovery activities, obstructing progress in other sectors.

Moreover, climate change has a profound impact on agriculture and food production. Rising temperatures, changing weather patterns, and altered precipitation levels can lead to reduced crop yields, increased pests and diseases, and disrupted ecosystems. These factors, combined, create uncertainty within the agricultural sector, leading to volatility in food prices. Farmers and agricultural workers bear the brunt of these consequences, often facing financial instability and reduced incomes.

Furthermore, climate change jeopardizes the planet's biodiversity, which has its own set of economic repercussions. Ecosystem services, such as pollination, water purification, and carbon sequestration, play a vital role in supporting economies worldwide. The loss of species and ecosystems, due to climate change, undermines these services, ultimately affecting industries and livelihoods that depend on them. For example, the decline of bee populations, essential for pollinating crops, threatens agricultural productivity and the profitability of farmers.

Additionally, the economic consequences of climate change extend beyond direct impacts to encompass indirect effects caused by policy responses. As governments recognize the urgency of addressing climate change, they implement regulations and policies to reduce greenhouse gas emissions. This transition to cleaner energy sources and greener technologies presents both opportunities and challenges for various sectors of the economy.

On one hand, investments in renewable energy and the development of new technologies can spur economic growth and job creation. Additionally, reducing dependence on fossil fuels can lead to an enhanced energy security and reduced healthcare costs due to improved air quality. However, industries heavily reliant on fossil fuels may experience challenges and face economic hardship during this transition, necessitating support and investment in retraining and job creation.

The economic consequences of climate change caused by fossil fuels, although complex, demand urgent attention. Mitigation efforts aimed at reducing greenhouse gas emissions and adapting to changing environmental conditions can help minimize these economic costs.

In conclusion, the economic implications of climate change caused by fossil fuels are vast and multifaceted. It affects several sectors, including infrastructure, agriculture, biodiversity, and energy. While the consequences entail substantial economic costs and challenges, embracing a sustainable and low-carbon future offers opportunities for economic growth and resilience. To mitigate the adverse economic effects of climate change, policymakers, businesses, and individuals must work hand-in-hand to transition to cleaner and more sustainable practices.

- Analysis of potential financial shifts in adapting to renewable energy sources

Analysis of Potential Financial Shifts in Adapting to Renewable Energy Sources

RENEWABLE ENERGY SOURCES, such as solar, wind, hydroelectric, and geothermal power, have gained significant traction in recent years. As the world faces growing concerns over climate change and depleting fossil fuel reserves, transitioning to renewable energy has become a crucial response. However, this transition necessitates a comprehensive analysis of the potential financial shifts that will be incurred during this process. This article explores the long-term financial implications of adapting to renewable energy sources, including the opportunities, challenges, and economic benefits associated with this transition.

Body:

1. Initial Costs:

One significant financial element of shifting to renewable energy is the initial investment required. While renewable energy installations can be expensive at first, technological advancements and economies of scale have dramatically reduced costs in recent years. By examining the potential for savings over time and the inevitable decline in fossil fuel availability, it becomes apparent that the initial costs associated with renewable energy conversions are offset in the long run.

2. Fuel Costs and Price Stability:

Fossil fuel-based energy production is heavily reliant on fuel prices, which are inherently volatile due to geopolitical and market fluctuations. Conversely, renewable energy sources leverage infinite natural resources (sun, wind, water), providing price stability and reducing exposure to volatile markets. This long-term price stability helps insulate consumers and businesses from sudden financial shocks and provides economic security.

3. Job Creation and Economic Growth:

The transition to renewable energy is a complete industry transformation, which brings opportunities for job creation and economic growth. New sectors emerge, such as solar panel manufacturing, wind turbine installation, and green construction, resulting in increased employment rates and local economic development. This shift offers positive financial implications by revitalizing communities and stimulating economic activity.

4. Incentives and Financial Support:

Governments worldwide are providing substantial incentives and financial support to promote renewable energy development. This includes tax credits, grants, and subsidies for installing renewable energy systems, further reducing the financial burden on individuals, businesses, and utilities. Additionally, renewable energy producers may sell excess energy to the grid, resulting in revenue generation and further offsetting their initial investments.

5. Enhanced Energy Independence:"

Renewable energy sources reduce reliance on fossil fuel imports, leading to greater energy independence for nations. This decreased dependence on foreign energy sources offers considerable financial benefits by mitigating price manipulations and geopolitical conflicts that frequently arise in energy markets. The potential to redirect funds that would otherwise be spent on importing fuel towards domestic investments can lead to economic growth and stability.

6. Resilience to Climate Change:

Mitigating climate change through the adoption of renewable energy sources presents a proactive pathway to financial stability. The frequency and intensity of climate-related disasters have dangerous economic ramifications. By embracing renewables, cities, regions, and countries can minimize these risks and potentially save billions in disaster relief and infrastructure repair costs, bolstering economic resilience and overall financial well-being.

THE FINANCIAL REALITIES of shifting to renewable energy sources are complex, but a thorough analysis highlights the vast potential for economic opportunities, stability, and long-term savings. By examining factors like initial costs, fuel price stability, job creation, incentives, energy independence, and resilience to climate change, it becomes evident that adopting renewables offers substantial positive shifts from both micro and macroeconomic standpoints.

This adaptive shift signifies a profound transition that not only addresses environmental concerns but also generates a win-win scenario for financial actors. By channeling financial resources towards sustainable energy systems, we can both safeguard our planet and create more resilient, prosperous economies for future generations.

Chapter 8: Technological Advances and Alternatives

In recent years, technological advancements have revolutionized various aspects of our lives. From healthcare to transportation and from communication to entertainment, these advances have not only improved the efficiency and convenience of our daily routines but have also opened up avenues for groundbreaking innovations. This chapter aims to delve into the myriad technological advances and alternatives that shape our present and hold immense potential for the future.

1. Artificial Intelligence (AI) and Machine Learning:

Artificial Intelligence (AI) and Machine Learning have emerged as the frontrunners in technology-driven advancements. AI, the simulation of human intelligence in machines, has received widespread attention due to its applications in diverse fields such as healthcare, finance, and transportation. Through advanced algorithms, AI systems can now analyze vast quantities of data, make complex decisions, and automate tasks previously reserved for humans. Machine Learning, a subset of AI, enables machines to learn from experience and improve their performance without explicit programming. This has paved the way for numerous groundbreaking developments, including autonomous vehicles, personalized medicine, and virtual assistants.

2. Internet of Things (IoT):

The Internet of Things has transformed the way we interact with everyday objects and devices. IoT refers to a network of physical objects (things) embedded with sensors, software, and other components to connect and exchange data with each other and the internet. This enables real-time monitoring, analysis, and control of various systems, resulting in increased efficiency and improved decision-making. From smart homes, wearables, and connected cars to industrial automation and intelligent cities, IoT integration has revolutionized multiple domains. Continual advancements in IoT technologies promise to enhance connectivity, automation, and data-driven insights, ultimately leading to a fully interconnected world.

3. Virtual Reality (VR) and Augmented Reality (AR):

Virtual Reality (VR) and Augmented Reality (AR) have opened up new realms of immersive experiences. VR replaces the real world with a simulated environment, enabling users to become fully immersed in digital content. This technology has found applications in gaming, entertainment, education, and even therapeutic interventions. On the other hand, AR overlays digital information onto the real world, enhancing our perception of reality. AR has seamlessly integrated with various industries, from retail and marketing to training and maintenance. As these technologies continue to evolve, they hold promise for transforming industries, amplifying human experiences, and shaping the future of entertainment and productivity.

4. Renewable Energy and Sustainable Solutions:

Technological advancements have also paved the way for alternative energy sources and sustainable solutions. With increasing concerns about climate change and depleting fossil fuels, renewable energy technologies such as solar, wind, and hydroelectric power have become crucial for a sustainable future. The advancements in these areas have led to improved efficiency, scalability, and affordability of renewable energy systems, making them increasingly viable alternatives to conventional energy sources. Moreover, innovative approaches like energy storage systems are addressing the intermittent nature of renewable energy, ensuring its reliability and accessibility even during periods of low generation.

TECHNOLOGICAL ADVANCEMENTS and alternatives continue to redefine our lives, enhancing productivity, connectivity, and sustainability. The era of AI and Machine Learning have given birth to autonomous systems while the growing network of IoT promises a fully interconnected future. VR and AR have revolutionized entertainment, and renewable energy technologies have provided hope for cleaner energy supplies. By understanding and embracing these technological advances and alternatives, we can actively participate in shaping a future that is efficient, sustainable, and filled with fascinating innovations.

- Overview of renewable energy solutions and their progress

Overview of Renewable Energy Solutions and Their Progress

RENEWABLE ENERGY SOURCES are being increasingly explored as alternatives to conventional fossil fuels due to their sustainable nature and potential for reducing the environmental impact of energy production. Shifting toward renewable energy not only promises a greener future but also offers opportunities for energy security, job creation, and economic growth. This overview seeks to highlight the main renewable energy solutions currently in use and their progress in terms of technological advancements, market deployment, and global adoption.

I. Solar Energy Solutions:

1. Photovoltaic (PV) Solar Power:

- Description: PV arrays convert sunlight directly into electricity through the photovoltaic effect.

- Technological Advances: Solar cell efficiency improvements, the advent of thin-film and flexible modules, and improved manufacturing processes.

- Progress: Expanding deployment globally, decreasing costs per watt, and increasing adoption in residential, commercial, and utility-scale applications.

2. Concentrated Solar Power (CSP):

- Description: CSP technologies concentrate solar energy to produce heat, which drives electricity generators.

- Technological Advances: Development of advanced concentrating systems, thermal storage solutions, and hybrid solar power plants.

- Progress: Major deployment in sunny regions, steady cost reduction, and growing integration of thermal storage for seamless power supply.

II. Wind Energy Solutions:

1. Onshore Wind Power:

- Description: Wind turbines convert wind energy into electricity through rotating blades.

- Technological Advances: Improved turbine design, larger rotor diameters, and increased offshore installation capacity.

- Progress: Rising global capacity, competitive costs, and efforts in optimizing wind farm layouts for greater efficiency.

2. Offshore Wind Power:

- Description: Wind turbines installed in marine environments, typically in shallow coastal waters or deeper offshore areas.

- Technological Advances: Advancements in floating turbine platforms, foundation structures, and cable systems for deep-sea installations.

- Progress: Rapid growth in capacity, decreasing costs, and ambitious offshore wind development, particularly in Europe.

III. Hydropower Solutions:

1. Conventional Hydropower:

- Description: Harnessing the potential energy of moving water to generate electricity through dam systems and turbines.

- Technological Advances: Improved turbine efficiency, eco-friendly fish passage systems, and optimization of reservoir operations.

- Progress: Established large-scale installations worldwide, expanding small-scale schemes, and integration with newer smart grid technologies.

2. Marine Energy (Wave and Tidal Power):

- Description: Harnessing energy from ocean waves and tides to generate electricity through various structural and converter technologies.

- Technological Advances: Enhanced power take-off systems, floating device designs, and undersea cable solutions.

- Progress: Marine energy still in early stages with several pilot projects and commercial-scale arrays undergoing testing and development.

IV. Bioenergy Solutions:

1. Biomass Power:

- Description: Utilizing biological resources such as agricultural products, forestry residues, or dedicated energy crops to produce heat and electricity.

- Technological Advances: More efficient combustion and gasification processes, advanced emission controls, and combined heat and power systems.

- Progress: Increasing use of biomass for heat and electricity generation, evolving conversion technologies, and increased research into sustainable feedstocks.

2. Biofuels:

- Description: Liquid fuels derived from biomass feedstocks (e.g., bioethanol from crops or municipal waste, biodiesel from vegetable oils).

- Technological Advances: Improved biofuel production methods, advanced feedstock cultivation, and algae-based biofuels.

- Progress: Biofuels being used as blends in transportation fuels, ongoing advancements in cellulosic ethanol production, and commercial algae biofuel ventures.

RENEWABLE ENERGY SOLUTIONS have made significant progress in recent years, driven by technological innovations, policy support, and growing environmental concerns. Solar and wind energy dominate as mature technologies, with decreasing costs and increasing deployment. Meanwhile, hydropower and bioenergy present vast untapped potential, and further investments and advances in marine energy hold promise for future utilization. By embracing renewable energy solutions, the world can accelerate its transition toward a cleaner, more sustainable energy future, reducing greenhouse gas emissions and mitigating climate change impacts.

- Key innovations in sustainable energy technologies and their feasibility

Key innovations in sustainable energy technologies have the potential to revolutionize the way we generate and consume energy, providing cleaner and more efficient alternatives to traditional fossil fuels. These innovations not only address the urgent need to reduce greenhouse gas emissions and mitigate the impacts of climate change, but also offer long-term economic benefits and energy security.

One major area of innovation is the development of renewable energy sources such as solar power, wind energy, and wave energy. Solar power has seen a remarkable advancement in recent years with the introduction of more efficient and cost-effective photovoltaic cells. These cells harness the energy from sunlight and convert it into electricity, offering a viable alternative to fossil fuel-based electricity generation. Solar thermal power plants, which use sunlight to heat a working fluid and generate steam, have also made significant progress, with improved efficiency and cost reductions.

Similarly, wind energy has witnessed significant investments and technological advancements. Modern wind turbines have taller towers and larger rotors, allowing them to capture more energy from the wind. Advanced control systems and improved blade designs have also addressed the intermittency and variability issues associated with wind power, making it a viable and reliable source of electricity.

Another promising innovation is the use of bioenergy, which involves utilizing organic materials such as crop residues, municipal waste, and purpose-grown crops to produce energy. This can be done through various methods such as anaerobic digestion, combustion, and gasification. Bioenergy offers significant potential for carbon reduction and can play a crucial role in waste management and agricultural sustainability.

Energy storage technologies are also key to enabling a reliable and efficient integration of renewable energy sources. Advancements in battery technologies, particularly lithium-ion batteries, have made them more cost-effective and efficient, allowing for increased energy storage on a grid-scale level. This enables the smooth integration and dispatchability of renewable energy to meet electricity demand, regardless of fluctuating renewable energy resources.

Furthermore, innovations in grid management and smart grid technologies are improving the overall efficiency of the power system. Smart grids utilize advanced communication infrastructure, real-time data analytics, and automation to optimize the transmission and distribution of electricity. This enables better demand response, efficient load balancing, and integration of decentralized energy generation, ultimately reducing waste and improving overall system resilience.

While these innovations hold substantial promise, their feasibility and widespread adoption require careful consideration of various factors. Cost plays a crucial role, as the initial capital investment for some sustainable energy technologies can be higher than conventional alternatives. However, with the projected technology advancements and economies of scale, the costs are expected to decrease, making them more economically viable.

Policy frameworks and government support are also instrumental in driving the adoption of sustainable energy technologies. Incentives such as tax credits, subsidies, and feed-in tariffs help decrease the financial burden and promote cleaner energy options. Additionally, research and development investments and collaboration between governments, academia, and the private sector are vital for pushing the boundaries of innovation and fostering progress in sustainable energy technologies.

Overall, the key innovations in sustainable energy technologies hold immense promise for a cleaner and more sustainable future. While their feasibility depends on factors such as cost, policy support, and technological advancements, continued research, development, and collaboration will undoubtedly drive their widespread adoption and ensure a smooth transition to a low carbon economy.

Chapter 9: International Agreements and Regulations

In our increasingly interconnected world, international agreements and regulations play a crucial role in shaping the global landscape. These agreements and regulations are designed to foster cooperation among nations, ensuring equitable development, protecting the environment, and promoting stability and peace. This chapter delves into the fascinating world of international agreements and their impact on various aspects of our lives.

1. Historical Context

To understand the significance of international agreements, we must first explore their historical context. The legacy of international cooperation can be traced back to ancient times, where treaties and agreements played a pivotal role in managing trade, resolving conflicts, and establishing diplomatic relations. However, the modern framework of international agreements was established after World War II, with the creation of institutions like the United Nations and regional organizations such as the European Union.

2. Types of International Agreements

International agreements can take various forms, ranging from bilateral treaties between two countries to multilateral conventions involving dozens or even hundreds of nations. These agreements can cover a wide array of issues, from trade and commerce to human rights and security. Some notable examples include the Paris Agreement on Climate Change, the Nuclear Non-Proliferation Treaty, and the Convention on the Rights of the Child.

3. Key Features of International Agreements

While the specifics may vary, most international agreements share certain key features. Firstly, they involve a mutual understanding and commitment among participating nations, often formulated through negotiations and diplomatic channels. Secondly, these agreements establish legal frameworks that bind signatory countries to adhere to certain standards and regulations. Finally, international agreements typically come with mechanisms for monitoring and enforcement to ensure compliance.

4. Global Economic Agreements

The world economy relies heavily on international agreements to facilitate trade and investment. Organizations such as the World Trade Organization (WTO) oversee the implementation of agreements such as the General Agreement on Tariffs and Trade (GATT), which aims to minimize trade barriers and promote fair global competition. Regional trade agreements, such as the North American Free Trade Agreement (NAFTA) and the Comprehensive and Progressive Agreement for Trans-Pacific Partnership (CPTPP), further enhance economic integration among participating nations.

5. Environmental Agreements

In recent decades, global awareness of environmental issues has led to the creation of several landmark agreements. The Paris Agreement, signed by nearly all countries, seeks to combat climate change through a collective effort to reduce greenhouse gas emissions. The Convention on Biological Diversity aims to conserve biodiversity while ensuring its sustainable use. Disposal of hazardous waste is regulated through agreements such as the Basel Convention. These environmental agreements demonstrate the commitment of nations to tackle pressing ecological challenges collectively.

6. Human Rights Agreements

Protecting and promoting human rights is a fundamental principle of international law. Numerous agreements address various aspects of human rights, such as civil and political rights, economic and social rights, and the rights of vulnerable groups. The International Covenant on Civil and Political Rights, the Universal Declaration of Human Rights, and the Convention on the Elimination of All Forms of Discrimination against Women are among the significant international treaties in this area, striving to safeguard human dignity across the globe.

7. Peace and Security Agreements

Ensuring global peace and security is a paramount objective of the international community. Agreements in this domain range from disarmament and non-proliferation treaties, such as the Nuclear Non-Proliferation Treaty, to conventions against terrorism and the illicit arms trade. Regional organizations, like the North Atlantic Treaty Organization (NATO), establish frameworks for collective defense and security cooperation among member states.

8. Challenges and Limitations

Despite their importance, international agreements face challenges in their implementation and enforcement. Differing national interests, geopolitical rivalries, and issues of compliance can hinder progress in achieving the desired outcomes. Additionally, negotiations can be complex and time-consuming, requiring compromises among participating nations. Addressing these challenges and ensuring effective implementation and enforcement remain crucial tasks for nations and international organizations.

INTERNATIONAL AGREEMENTS and regulations are vital mechanisms for addressing global challenges and promoting harmonious cooperation between nations. They encompass a broad range of issues, from economic cooperation and environmental protection to human rights and security. As our world becomes increasingly interconnected, fostering a strong framework of international agreements is essential to create a more just, prosperous, and sustainable future for all.

- Evaluation of global efforts to address global warming, such as the Paris Agreement

The evaluation of global efforts to address global warming, particularly in relation to the Paris Agreement, is essential to assess the effectiveness and progress made in reducing greenhouse gas emissions and combating climate change. This article aims to provide a detailed and interesting analysis of these efforts, discussing the key elements of the Paris Agreement and examining the achievements, challenges, and potential outcomes derived from global attempts to tackle global warming.

The Paris Agreement, adopted in December 2015, is a comprehensive international agreement aiming to limit global warming to well below 2 degrees Celsius above pre-industrial levels, pursuing efforts to limit the temperature increase even further to 1.5 degrees Celsius. It entered into force in November 2016, marking a significant milestone in global efforts to confront climate change.

One of the central strengths of the Paris Agreement is its universal nature, with 196 Parties signing and 190 ratifying the accord thus far. This widespread participation ensures broad support and commitment to addressing climate change and provides a platform for collaborative action at the global level. The agreement also promotes a bottom-up approach, allowing countries to determine their own contributions, known as Nationally Determined Contributions (NDCs), to emissions reductions. This flexibility recognizes the differing circumstances, capabilities, and responsibilities of each country and encourages them to pursue their climate goals with aligned efforts.

Since the inception of the Paris Agreement, many notable achievements have been made. The growth of renewable energy sources has created a more sustainable energy landscape, with countries transitioning from fossil fuel-dependent economies to cleaner alternatives. Technological advancements and falling costs have contributed to the increased adoption of renewable energy, leading to significant emissions reductions in several regions worldwide.

Furthermore, global coalitions such as the International Solar Alliance and Mission Innovation have been instrumental in fostering international cooperation and knowledge-sharing to further accelerate the transition to clean energy. These collaborative efforts are essential in leveraging resources and expertise to develop innovative solutions that can address the challenges posed by climate change effectively.

However, it is important to acknowledge the challenges and potential drawbacks of the Paris Agreement. Firstly, the agreement's voluntary nature means that there are no binding regulations or penalties for non-compliance. This can somewhat undermine the agreement's enforceability and create disparities in ambition levels among participating countries. Additionally, the Paris Agreement relies heavily on individual countries' commitment to voluntarily implement emissions reduction strategies, which may vary widely in terms of ambition and effectiveness.

Another significant challenge is the financing aspect of global efforts to address climate change. Developed countries have pledged to provide financial support to developing countries to assist them in mitigating and adapting to climate change. However, there is still a gap between the funds pledged and those actually provided, hindering the progress of vulnerable nations in their climate action endeavors.

Looking ahead, it is evident that much progress still needs to be made to accelerate the pace of emissions reductions and limit global warming to the desired targets. Countries must continuously review and strengthen their NDCs to enhance the collective ambition, setting new and more ambitious targets to align with the updated science and the evolving energy landscape.

The current global crisis created by the COVID-19 pandemic brings both challenges and opportunities for global efforts to address global warming. The crisis has led to a temporary reduction in global emissions due to economic slowdowns and lockdown measures, offering a glimpse of what might be achieved through profound changes in our lifestyles and energy systems. It is essential to seize this moment and implement sustainable recovery plans that align with climate objectives to ensure a resilient and low-carbon future.

In conclusion, the evaluation of global efforts to address global warming, specifically the Paris Agreement, provides valuable insights into the achievements, challenges, and potential outcomes of international collaboration in combating climate change. While progress has been made in emissions reductions and the transition to clean energy, there are still hurdles to overcome, including enforcement issues, varying levels of ambition, and funding gaps. Continually reassessing and strengthening commitments, fostering international cooperation, and seizing opportunities for sustainable recovery are key to tackling global warming and securing a vibrant and sustainable future for all.

- Examination of regulations placed on the fossil fuel industry worldwide

The fossil fuel industry plays a crucial role in powering the global economy. However, its activities have also raised concerns about environmental degradation, climate change, and public health. As a result, governments worldwide have implemented a variety of regulations aimed at mitigating these issues and ensuring sustainable practices within the industry. This examination will dive into the regulations placed on the fossil fuel industry worldwide, exploring their extent, effectiveness, and potential for improvement.

Starting with North America, regulations in the United States vary significantly across different levels of government. At the federal level, one of the most notable regulations is the Clean Air Act, enacted in 1970 and later amended to cover greenhouse gas emissions. This includes requirements for emission controls and the establishment of emission standards for major pollutants like sulfur dioxide, nitrogen oxides, and particulate matter. The Environmental Protection Agency (EPA) oversees the enforcement of these regulations.

Additionally, the U.S. has seen increased regulations at the state level. For instance, California has implemented its own comprehensive regulations to reduce greenhouse gas emissions. This includes the Global Warming Solutions Act of 2006, which aims to reduce greenhouse gas emissions to 1990 levels by 2020 through various measures like a cap-and-trade system.

Moving to another North American country, Canada has also implemented various regulations on the fossil fuel industry. The country has an extensive framework of federal, provincial, and territorial regulations that address environmental and climate-related concerns. The Canadian Environmental Assessment Act, for example, requires environmental assessments for major energy projects, ensuring potential risks and impacts are thoroughly evaluated.

Turning to Europe, the European Union (EU) has positioned itself as a global leader in climate action. The EU Emissions Trading System (EU ETS), launched in 2005, is one of the most significant regulations. It sets an emissions cap for various industries, including fossil fuel power stations and refineries, and provides a market for trading emissions allowances, encouraging emissions reduction across member states.

Furthermore, the EU has implemented various regulations targeting the phase-out of coal as a fossil fuel. The Coal Phase-Out Directive, adopted in 2017, supports member states in defining strategies to transition away from coal by setting emissions limits and requiring the application of best available techniques.

Outside of North America and Europe, countries like China and India have also introduced regulations to address environmental concerns in the fossil fuel industry. China, the world's largest carbon emitter, has implemented measures like the Clean Air Action Plan and the Energy Development Strategy Action Plan to combat air pollution and to promote energy efficiency and renewable energy usage.

Similarly, India has undertaken initiatives to improve air quality and fuel standards. Its national emission standards set limits for pollutants from different sources, including the fossil fuel industry. The gradual shift toward cleaner fuels and stricter regulations is evident in India's implementation of a national clean air program and the push for renewable energy development.

While significant progress has been made by governments worldwide in regulating the fossil fuel industry, challenges remain. Enforcement and compliance, for instance, are always areas of concern. Without strict oversight and monitoring, regulations may not yield the desired outcomes. Additionally, some argue that regulatory loopholes or leniencies can allow new environmental risks to emerge.

Moreover, the rapid pace of global energy consumption necessitates continuous improvement of regulations. Governments should adapt to new technological advancements, explore alternative energy sources like renewables, and incentivize industry innovation to reduce greenhouse gas emissions.

In conclusion, the examination of regulations placed on the fossil fuel industry worldwide highlights the varying approaches taken by different countries and regions. While regulations vary in their specifics and impact, they all aim to address environmental degradation, climate change, and public health concerns associated with fossil fuel extraction and consumption.

Governments must continue to enforce stricter regulations, ensure compliance, and promote sustainable practices as we strive for a cleaner and greener future.

62

Chapter 10: Societal Movements and Activism

In recent years, the world has witnessed an upsurge in societal movements and activism. From calls for racial justice to environmental activism and women's liberation movements, individuals and communities all over the globe are coming together to fight for change. In this chapter, we will delve into the intricacies of these movements and explore their impact on society.

1. The Nature of Societal Movements:

Societal movements can be defined as collective efforts by individuals who share a common goal and mobilize to bring about social and political change. These movements often challenge established power structures, policies, and norms that perpetuate inequality and injustice. They arise from a shared dissatisfaction with society's status quo and seek transformative change.

2. Historical Examples of Societal Movements:

Throughout history, numerous societal movements have left a lasting impact on society and paved the way for future activism. The civil rights movement in the United States during the 1960s fought against racial segregation and won significant victories, including the desegregation of schools and the passage of the Civil Rights Act of 1964. Similarly, the feminist movement fought for gender equality and women's rights, leading to changes in legislation, such as the right to vote and reproductive rights.

3. Current Societal Movements:

Today, several societal movements have gained traction and are at the forefront of global activism. The Black Lives Matter movement, sparked by the killings of unarmed Black individuals by the police, has brought systemic racism and police brutality into the global spotlight. Youth-led movements like Fridays for Future, sparked by Greta Thunberg's school strikes, have mobilized millions of young people to demand action on climate change. These movements highlight the power of collective action and the force of grassroots mobilization.

4. Intersectionality and Social Movements:

An important concept in societal movements is intersectionality, which recognizes that individuals can face multiple forms of oppression based on their social identities. Movements like Black Feminism and LGBTQ+ activism have emerged to address the unique challenges faced by individuals at the intersection of race, gender, and sexuality. Intersectional movements aim to create a more inclusive and equitable society that acknowledges and fights against all forms of oppression.

5. Online Activism and Digital Technology:

The advent of social media and digital technology has transformed the landscape of activism. Movements now have wider platforms to raise awareness, connect with like-minded individuals, and organize campaigns. Hashtags like #MeToo and #AintNoCinderella have gone viral, initiating conversations about sexual harassment and women's safety in different parts of the world. However, there are also criticisms that online activism does not always lead to tangible change and may overlook the importance of grassroots organizing.

6. Impact and Challenges of Societal Movements:

Societal movements have undeniably brought about tangible changes in legislation, public discourse, and social attitudes. They have elevated marginalized voices and forced societies to confront long-standing injustices. However, movements also face challenges, including backlash from those resistant to change, co-option by mainstream politics, and divisions within the movement itself. Movements must navigate these obstacles while maintaining a focus on their goals to ensure sustained progress.

SOCIETAL MOVEMENTS and activism play a crucial role in changing societies for the better. They challenge the existing structures of power, fight against various forms of injustice, and create spaces for marginalized voices to be heard. By understanding the nature of these movements, historical examples, current struggles, and the impact of digital technology, we can appreciate the importance of these movements in shaping our collective future.

- Discussion on the rise of grassroots movements advocating for sustainable change

In recent years, there has been a significant rise in grassroots movements advocating for sustainable change. These movements, driven by concerned citizens throughout the world, have emerged as a response to the pressing environmental and social issues that our planet currently faces. From climate change and deforestation to social inequality and injustice, these grassroots initiatives are galvanizing individuals and communities to take action, demanding that governments and corporations prioritize sustainability and commit to meaningful change.

One of the key factors behind the rise of grassroots movements advocating for sustainability is the growing awareness of the dire consequences of environmental degradation. With scientific evidence increasingly pointing towards the severity of global warming and the depletion of natural resources, more people are recognizing the urgent need for sustainable practices. Previously overlooked issues, such as deforestation and plastics pollution, have now become focal points of public attention. Grassroots movements have filled the void left by governments and corporations in addressing these problems, mobilizing individuals to raise their voices and demand change.

Social media has played a crucial role in this rise of grassroots movements. Platforms like Facebook, Twitter, and Instagram have given individuals an unprecedented reach and ability to organize and promote their causes. From hashtags and viral campaigns to petitions and online events, social media has empowered grassroots movements to galvanize large numbers of people, transcending geographical and cultural boundaries. This has resulted in a wider dissemination and accelerated growth of sustainable ideas and initiatives.

Moreover, the rise of grassroots movements can also be attributed to a growing disillusionment with current political and economic systems. As trust in governments and large corporations declines, people are increasingly turning to grassroots initiatives as viable alternatives. These movements offer an opportunity for individuals to directly engage in creating change rather than placing their trust solely on institutions that are perceived as unresponsive or corrupt. These grassroots movements allow people to take matters into their own hands, fostering a sense of empowerment and ownership over the change they wish to see in the world.

Additionally, grassroots movements advocating for sustainability often adopt a bottom-up approach, focusing on local communities and immediate issues. This localized action is more accessible and tangible for individuals, making it easier for them to engage and have a meaningful impact. By addressing local environmental and social issues, grassroots movements not only create a catalyst for change but also strengthen the fabric of communities. Through education, awareness campaigns, and sustainable practices at the grassroots level, individuals become active contributors to sustainable development, inspiring others to follow suit.

While grassroots movements may be seen as small-scale, they have the potential to create cascading effects that can influence larger institutions and policies. As these movements gain momentum and public support, their demands for sustainability permeate political discourse, putting pressure on decision-makers to take action. Organizations like Extinction Rebellion and Fridays for Future have successfully harnessed public attention and sparked a global conversation on climate change. In turn, governments and corporations have been forced to respond, acknowledging the concerns raised by grassroots movements and taking steps to address them.

In conclusion, the rise of grassroots movements advocating for sustainable change can be attributed to a combination of factors. Increased awareness of environmental issues, the power of social media, disillusionment with current systems, and a localized approach all contribute to the growth and success of these movements. They offer a platform for individuals to take meaningful action, demanding that sustainability be prioritized at both the grassroots and institutional levels. As more people recognize the importance of grassroots movements, their impact on achieving a more sustainable future is likely to grow.

- Exploration of the impact of societal pressure on the fossil fuel industry

Exploration of the Impact of Societal Pressure on the Fossil Fuel Industry

THE FOSSIL FUEL INDUSTRY has long been emblematic of economic progress and energy production. However, in recent years, significant societal pressure has emerged, demanding a shift toward cleaner and more sustainable energy sources. These societal pressures, driven by concerns such as climate change and environmental degradation, have significantly impacted the fossil fuel industry around the world. This essay aims to explore the multifaceted impact of societal pressure on the fossil fuel industry, highlighting various aspects such as regulatory measures, public sentiment, and industry responses.

1. Societal Pressure as a Catalyst for Regulatory Measures:

Societal concerns over the environmental consequences of fossil fuel extraction and combustion have prompted policymakers worldwide to introduce stringent regulations. Governments have enacted policies aimed at reducing greenhouse gas emissions, encouraging renewable energy adoption, and limiting fossil fuel exploitation. These measures, such as carbon taxes, emission caps, and renewable energy quotas, indirectly push fossil fuel companies to recalibrate their operations or seek alternative energy resources.

2. Public Sentiment and Consumer Actions:

Public sentiment is a driving force behind societal pressure. Increased awareness of the greenhouse gas emissions associated with fossil fuels has fostered a greater focus on sustainability. This awareness, combined with advancements in communication technology, has amplified the public's ability to express discontent and call for change. Consumers are increasingly exercising their power by favoring environmentally friendly products and companies. As a result, fossil fuel-dependent industries are facing mounting challenges to their products' marketability and reputation.

3. Divestment Movements:

The fiercely debated divestment movement is another manifestation of societal pressure on the fossil fuel industry. Activists and socially conscious investors are putting pressure on universities, pension funds, and other financial institutions to divest from fossil fuels. Such commitments can be detrimental to the financial stability of fossil fuel companies, eroding investor confidence and lowering access to funding. Consequently, the divestment movement contributes to an already challenging operating environment for fossil fuel extraction and exploration.

4. Investment Shifts:

Societal pressure has caused a shift in investment patterns, as portfolios increasingly prioritize renewable energy sources over fossil fuel projects. This reallocation of funds stems from a combination of ethical considerations, market dynamics, and decarbonization imperatives. Growing investment in renewables provides economic alternatives while limiting the industry's potential profitability, thus reinforcing societal concerns surrounding fossil fuels.

5. Corporations' Response:

Faced with mounting pressure, fossil fuel companies have been forced to adapt or risk becoming obsolete. Industry giants are diversifying their portfolios by investing in cleaner energy technologies and supporting research and development efforts. Some companies have committed to reducing emissions, transitioning to a low-carbon future, and aligning their business practices with sustainability goals. However, differences in the scale and sufficiency of such measures persist, leading to enduring opposition from activist groups.

SOCIETAL PRESSURE HAS had a profound impact on the fossil fuel industry, catalyzing regulatory actions and spurring innovative approaches within the sector. The combination of public sentiment, divestment movements, shifts in investment patterns, and climate-friendly corporate initiatives has altered the landscape of traditional energy production. As the push for renewable energy sources gains momentum, it is imperative for fossil fuel companies to engage in sustainable practices, embrace new technologies, and seek synergies between profitability and environmental responsibility. The interplay between societal pressure and the fossil fuel industry will continue to be a critical factor in fostering a global transition toward a cleaner and greener energy paradigm.

Chapter 11: Environmental Impacts of Fossil Fuels Beyond Global Warming

The burning of fossil fuels, such as coal, oil, and natural gas, has been the backbone of our modern industrialized society. However, it is well-known that the combustion of these fuels is a major contributor to global warming and climate change. The release of greenhouse gases, especially carbon dioxide, into the atmosphere traps heat and leads to an increase in the Earth's average temperature. This chapter, on the other hand, delves into other often-overlooked environmental impacts associated with fossil fuel consumption. These impacts go beyond global warming and provide a broader perspective on the harmful consequences of our reliance on fossil fuels.

1. Air Pollution:

Fossil fuel combustion releases a plethora of pollutants into the air that pose significant health risks. These pollutants include sulfur dioxide (SO_2), nitrogen oxides (NO_x), particulate matter (PM), volatile organic compounds (VOCs), and mercury. Acid rain, smog formation, respiratory illnesses, and cardiovascular diseases are just a few examples of the detrimental effects of air pollution resulting from burning fossil fuels. In fact, according to the World Health Organization (WHO), outdoor air pollution from fossil fuel pollution leads to the premature deaths of approximately 4.6 million people each year.

2. Water Pollution:

The extraction, production, and transportation of fossil fuels can contribute to water pollution. Oil spills from offshore drilling or transport accidents are particularly notorious, as they cause devastating ecological damage and harm marine organisms, affecting entire ecosystems for years, if not decades. Moreover, the wastewater generated from fracking, a technique used to extract natural gas, contains toxic chemicals that can contaminate groundwater and pollute surface water sources.

3. Land Degradation:

Fossil fuel extraction often leads to direct and indirect land degradation. Surface mining for coal, for example, involves the removal of vast areas of land, stripping away topsoil, and destroying vegetation. This results in the loss of biodiversity, disrupting ecosystems, and rendering the land nearly incapable of supporting future growth. Similarly, oil and gas extraction can also cause habitat destruction, fragmentation, and soil contamination. Large-scale industrial facilities such as refineries and power plants further contribute to land degradation through infrastructural development.

4. Deforestation:

In order to extract fossil fuels, vast expanses of forests are often cleared to set up drilling sites or create mining areas. This deforestation eliminates critical habitats for numerous plant and animal species, leading to biodiversity loss. Forests also act as carbon sinks, absorbing large amounts of CO2 from the atmosphere. Therefore, deforestation resulting from fossil fuel extraction not only releases carbon dioxide stored in trees but also diminishes the planet's capacity to sequester more CO2, exacerbating climate change.

5. Habitat Destruction and Wildlife Displacement:

The construction and operation of fossil fuel infrastructure directly impact wildlife by destroying their habitats, fragmenting ecosystems, and displacing species. Pipelines, power lines, and roads cut through critical habitats, posing significant barriers for migration and disrupting natural corridors. Additionally, the noise, light, and air pollution associated with fossil fuel extraction and transportation further disturb and displace wildlife populations, jeopardizing their survival and ecological balance.

6. Coal Ash and Mining Waste:

The burning of coal produces a significant amount of waste in the form of coal ash. If not managed properly, coal ash can leach heavy metals and other toxic substances into the environment, contaminating groundwater and surface water sources. Likewise, mining waste, known as tailings, which is produced during various stages of extraction, can pollute nearby water bodies, causing long-term ecological damage and threatening aquatic ecosystems.

WHILE GLOBAL WARMING and climate change are the most commonly recognized environmental impacts associated with fossil fuel consumption, it is crucial to understand the broader spectrum of ecological damage caused by these fuels. From air and water pollution to land degradation, deforestation, and wildlife displacement, the environmental impacts of fossil fuels reach beyond the single issue of climate change. It is imperative that society collectively acknowledges these adverse effects and actively pursues sustainable alternatives to reduce our reliance on fossil fuels and mitigate their damaging impacts.

- Analysis of other ecological consequences caused by the industry

The industry has had numerous ecological consequences that are not commonly discussed. While most people are aware of issues such as pollution and deforestation, there are several other impacts that often go unnoticed.

One major consequence of the industry is habitat destruction. As companies clear land to make room for operations and infrastructure, countless species lose their homes. This loss of habitat can lead to a decrease in biodiversity and even the extinction of certain species. Moreover, when habitats are destroyed, it disrupts the delicate balance of ecosystems, causing ripple effects throughout the food chain.

Another ecological consequence of the industry is the alteration of natural water systems. From mining operations to industrial agriculture, activities within the industry can have a detrimental impact on rivers, lakes, and groundwater. Chemical runoff from farms can contaminate water sources, making them unfit for consumption and causing harm to aquatic life. Additionally, extractive industries often require large amounts of water, leading to the depletion of freshwater sources in many areas.

Air pollution is another significant consequence of the industry. Emissions from factories, power plants, and vehicles contribute to the release of harmful pollutants into the atmosphere. These pollutants can have various adverse effects on both human health and the environment. For instance, they can lead to respiratory problems, smog formation, and the depletion of the ozone layer, which in turn, exacerbates climate change.

Speaking of climate change, the industry is also a significant contributor to greenhouse gas emissions. Fossil fuel extraction, energy production, and transportation all release vast amounts of carbon dioxide and other greenhouse gases, leading to global warming. The continuous extraction and burning of these fossil fuels are accelerating the warming of the planet and causing extreme weather events, rising sea levels, and the decline of fragile ecosystems such as coral reefs.

Furthermore, the industry has a significant impact on land and soil degradation. Activities like mining and logging often leave landscapes barren and unproductive. Over time, these degraded lands become vulnerable to erosion and desertification, making them unsuitable for vegetation growth. As a result, soil quality decreases, affecting agricultural productivity and exacerbating food insecurity in many regions.

Lastly, the industry has also had adverse effects on wildlife populations. Hunting and poaching for various purposes, from animal skins to traditional medicine, have caused significant declines in certain species. Moreover, illegal wildlife trade feeds into the profits of criminal networks and further threatens endangered species, pushing them closer towards extinction.

In conclusion, the ecological consequences caused by the industry are far-reaching and must be carefully addressed. Habitat destruction, alteration of water systems, air pollution, climate change, land and soil degradation, and the decline of wildlife populations are just some of the many repercussions of industrial activities. It is imperative that we recognize and prioritize the mitigation of these consequences in order to ensure a sustainable future for our planet.

- Examination of pollution, deforestation, and habitat destruction

Examination of Pollution, Deforestation, and Habitat Destruction

Pollution, deforestation, and habitat destruction are significant environmental issues that affect our planet in many ways. These processes pose serious threats to the Earth's biodiversity and the overall health of ecosystems. In this article, we will delve into the intricacies of these topics, exploring their causes, impacts, and potential solutions.

Let us begin with pollution- the introduction of harmful substances or pollutants into the environment. Pollution can be caused by various activities, including industrial processes, transportation, agricultural practices, and improper waste disposal. The most common types of pollution include air pollution, water pollution, and soil pollution.

Air pollution results from the release of pollutants into the atmosphere, primarily from human activities such as burning fossil fuels, industrial emissions, and the use of vehicles. This can lead to detrimental effects on human health, such as respiratory problems, as well as harm to plant and animal life. Additionally, air pollution contributes to the depletion of the ozone layer and climate change.

Water pollution is caused by the discharge of pollutants into lakes, rivers, oceans, and other water bodies. These pollutants can come from industrial sources, agricultural runoff, sewage discharge, or even chemical spills. Water pollution not only affects the quality of drinking water but also harms aquatic ecosystems and their inhabitants. Marine life can suffer from reduced oxygen levels, toxic chemical exposures, and habitat degradation.

The third type of pollution is soil pollution, also known as land pollution. Soil pollution occurs when toxic chemicals, heavy metals, or waste materials contaminate the soil. Pesticide and fertilizer misuse, industrial waste disposal, improper sewage treatment, and mining activities are all significant contributors to this form of pollution. Soil pollution affects the fertility of the land, posing direct threats to agriculture and food security.

Deforestation, the second environmental issue we will now explore, involves the removal of forested areas for non-forestry purposes. This process is primarily driven by urbanization, commercial logging, agricultural expansion, and the extraction of natural resources. Forests serve as crucial habitats for numerous plant and animal species, and their destruction leads to the loss of biodiversity, disrupting essential ecological balance.

Deforestation causes a myriad of other problems as well. It contributes significantly to global greenhouse gas emissions and climate change as trees, when burned or removed, release stored carbon dioxide into the atmosphere. Deforestation also leads to soil erosion, negatively impacting nutrient cycling and contributing to water pollution due to sediment runoff. It can even disrupt local weather patterns and harm indigenous societies reliant on forest resources, such as indigenous peoples or local communities.

Lastly, habitat destruction refers to the broad process of environmental degradation that results in the loss or severe alteration of habitats. Apart from deforestation, which is a major driver of habitat destruction, urbanization, infrastructure development, and unsustainable agricultural practices also play significant roles. As habitats vanish, many species cannot adapt or find suitable alternatives, leading to declines in populations and potential extinction.

The destruction of habitats leads to unbalanced ecosystems, negatively impacting biodiversity, and hampering ecosystem services. Ecosystem services are benevolent aspects that nature provides, such as water purification, pollination, and carbon sequestration, which are essential for human well-being and sustainable development. A loss of habitat also directly affects indigenous cultures and communities that depend on the land and its resources for their livelihoods and cultural practices.

While the issues of pollution, deforestation, and habitat destruction appear daunting, numerous solutions and strategies can address these concerns. Governments and organizations worldwide are actively promoting sustainable practices such as implementing stricter environmental regulations, decreasing reliance on fossil fuels, and promoting renewable energy sources. Afforestation and reforestation efforts are also being emphasized to restore and expand forest cover.

Similarly, promoting eco-friendly agricultural practices, reducing industrial pollution through advanced technologies, and improving waste management systems are crucial steps in combatting pollution. Additionally, raising awareness among individuals through education and encouraging conscious consumerism can help create a more environmentally conscious society.

In conclusion, pollution, deforestation, and habitat destruction pose significant challenges to the Earth's ecosystems and its inhabitants. Understanding the causes, impacts, and solutions to these issues is vital for improving environmental sustainability and ensuring a livable planet for future generations. By taking collective action and implementing environmentally-friendly practices, we can mitigate these problems and work towards a healthier and more harmonious coexistence with nature.

Chapter 12: Adaptation and Mitigation Strategies

In the face of climate change, there are two key strategies that can be employed: adaptation and mitigation. While both are essential in addressing the challenges posed by a changing climate, they differ in their approach and goals. In this chapter, we will delve into the intricacies of these strategies, exploring the potential benefits and limitations they offer.

Adaptation refers to the process of adjusting to the impacts and consequences of climate change. It involves making changes and implementing measures to minimize vulnerability and enhance resilience in various systems, such as ecosystems, communities, and economies. Adaptation strategies aim to reduce the negative impacts of climate change and take advantage of any potential opportunities that may arise.

One of the most crucial aspects of adaptation is assessing vulnerability. By identifying the areas, sectors, or populations most susceptible to climate change impacts, policymakers and stakeholders can develop targeted and effective adaptation strategies. Vulnerability assessments should take into account factors such as exposure, sensitivity, and adaptive capacity. For example, coastal communities are often highly vulnerable due to their exposure to sea-level rise and hurricanes, while developing countries may have limited financial resources for adaptation measures.

Various adaptation strategies can be employed depending on the system and the level of vulnerability. In natural ecosystems, for instance, adaptation may involve habitat restoration, species reintroduction, or the creation of artificial habitats. These measures can help preserve biodiversity and ensure ecosystem services are maintained, even under changing climate conditions.

In human communities, adaptation strategies can range from improved infrastructure and land-use planning to the development of early warning systems and emergency response plans. For vulnerable groups, such as the elderly or low-income populations, social adaptation measures, such as improved healthcare and social support systems, are crucial to ensure their resilience to climate change impacts.

While adaptation is crucial, it alone cannot solve the problem of climate change. This is where mitigation strategies come into play. Mitigation aims to reduce greenhouse gas emissions and prevent further climate change by minimizing the sources or enhancing the sinks of these gases. Unlike adaptation, which deals with managing the impacts of climate change, mitigation addresses the root cause.

One of the most effective mitigation strategies is the reduction of fossil fuel use. Transitioning to cleaner and renewable energy sources, such as solar, wind, and hydroelectric power, can significantly decrease greenhouse gas emissions associated with energy production. Additionally, improving energy efficiency in buildings, transportation, and industrial processes can make a substantial contribution to greenhouse gas reduction efforts.

Another important aspect of mitigation is protecting and enhancing carbon sinks, such as forests and oceans, as they play a crucial role in absorbing carbon dioxide from the atmosphere. Deforestation and land degradation not only contribute to emissions but also diminish the planet's capacity to absorb greenhouse gases naturally. Therefore, reforestation, afforestation, and sustainable land management practices are vital for mitigating climate change.

Furthermore, technological advancements and innovation can play a significant role in mitigating climate change. For instance, the development and deployment of carbon capture and storage technologies can help reduce emissions from fossil fuel power plants and industrial processes. Additionally, supporting research and development efforts in clean energy technologies can accelerate the transition to a low-carbon economy.

While adaptation and mitigation are distinct approaches, they are not mutually exclusive. In fact, integrated strategies that encompass both adaptation and mitigation can yield enhanced benefits and be more efficient in addressing the challenges of climate change. For example, implementing adaptation measures that also contribute to emissions reductions, such as green infrastructure or sustainable agriculture practices, can optimize resources and achieve both adaptation and mitigation goals concurrently.

In conclusion, adaptation and mitigation are two essential strategies in addressing climate change. Adaptation focuses on adjusting to the impacts and reducing vulnerability, while mitigation involves reducing greenhouse gas emissions and tackling the root causes of climate change. Both strategies are vital in building resilience, protecting ecosystems, and securing human well-being in the face of an increasingly unpredictable climate. By implementing integrated approaches that consider both adaptation and mitigation, societies can take significant strides towards a sustainable and climate-resilient future.

- Overview of efforts to adapt to and mitigate the effects of global warming

Global warming, caused by the increase in greenhouse gases in the Earth's atmosphere, is one of the most pressing challenges that humanity faces today. The consequences of ongoing global warming are far-reaching and include rising sea levels, extreme weather events, and the loss of natural habitats and biodiversity. In order to combat these effects and prevent further damage to the planet, countries around the world have been implementing various strategies and initiatives to adapt to and mitigate global warming.

One of the key areas of focus in adapting to the effects of global warming is the implementation of climate change resilience measures. This involves building infrastructure and systems capable of withstanding the impacts of climate change. For example, constructing flood defenses and bolstering coastal erosion protection can help mitigate the damage caused by rising sea levels and increased storm surges. Additionally, redesigning urban areas to be more resilient to extreme heat events and ensuring reliable access to water resources are crucial adaptations for urban centers.

Another important aspect of adapting to global warming is sustainable land management practices. This includes reforestation projects, promoting sustainable agriculture techniques, and combating deforestation. Reforestation schemes not only help sequester carbon dioxide, but they also provide numerous co-benefits such as the creation of wildlife habitats and the prevention of soil erosion. By adopting sustainable agricultural practices, such as organic farming and agroforestry, farmers can reduce greenhouse gas emissions from agricultural activities while maintaining crop yields.

Furthermore, efforts to adapt to global warming must also involve the protection and restoration of ecosystems, such as forests, wetlands, and coral reefs. These ecosystems act as natural carbon sinks and are essential for preserving wildlife and biodiversity. Support for initiatives like protected-area management, creating marine sanctuaries, and implementing sustainable fishing practices are crucial to conserving these valuable resources.

In addition to adaptation efforts, global cooperation is also required to mitigate the effects of global warming. The primary approach to mitigation involves reducing greenhouse gas emissions. This can be achieved through a combination of transitioning to renewable energy sources, improving energy efficiency, and implementing sustainable transport systems. Countries are increasingly investing in renewable energy infrastructure, such as solar and wind farms, to replace fossil fuel-based power plants. Energy efficiency measures, such as improved insulation in buildings and the use of energy-efficient appliances, are also being implemented to reduce energy consumption.

Furthermore, promoting sustainable transport options, such as public transportation and electric vehicles, can significantly reduce greenhouse gas emissions from the transportation sector. Additionally, carbon capture and storage technologies are being developed to capture and store carbon dioxide emitted from fossil fuel-based industries, such as power plants and cement production.

Overall, the efforts to adapt to and mitigate the effects of global warming are multifaceted and require a combination of strategies at the global, national, and local levels. This includes implementing resilience measures, adopting sustainable land management practices, protecting and restoring ecosystems, and reducing greenhouse gas emissions through transitioning to renewable energy and improving energy efficiency. By working together to address global warming, we can create a sustainable future for generations to come.

- Role of governments, organizations, and individuals in implementing change

Role of Governments in Implementing Change:

1. Policy Formation: Governments play a crucial role in implementing change by formulating and implementing policies that address societal challenges. These policies set the direction and framework for change. For instance, governments may develop policies to combat climate change, promote sustainable development, or improve healthcare access.

2. Regulatory Framework: Governments establish laws, regulations, and standards to guide and enforce the desired changes. These regulations ensure compliance and provide guidelines for businesses, organizations, and individuals to adopt new practices. For example, environmental regulations may mandate emission reduction targets for industries to mitigate pollution.

3. Resource Allocation: Governments have the power to allocate resources and funding to support the implementation of change. They can invest in research and development, infrastructure, and capacity-building initiatives. Governments may also offer financial incentives, grants, or tax breaks to encourage individuals and organizations to adopt desired changes.

4. International Collaboration: Governments play a role in bringing about global change by collaborating with other nations to address transnational issues. They negotiate and ratify international agreements to promote cooperation and drive change on a global scale. These agreements may focus on areas such as human rights, trade, or environmental conservation.

Role of Organizations in Implementing Change:

1. Research and Innovation: Organizations, both private and public, contribute to change implementation by investing in research and innovation. They develop new technologies, practices, and methodologies that can bring about desired changes. Research organizations can identify problems, analyze data, and propose evidence-based solutions.

2. Advocacy and Awareness: Organizations have the ability to raise awareness about the need for change and advocate for specific causes. They

can use their platforms to inform and educate the public, create movements, and mobilize support. NGOs, interest groups, and professional organizations often play active roles in advocating for changes in policies, laws, and societal behaviors.

3. Pilot Projects and Demonstrations: Organizations can test and implement change on a smaller scale through pilot projects and demonstrations. These initiatives allow organizations to assess the feasibility and effectiveness of new practices, policies, or technologies before broader implementation. Lessons learned from these projects can inform the larger-scale change implementation.

4. Capacity-Building: Organizations can facilitate change by building the capacity of individuals and communities to adopt new practices. They provide training, knowledge-sharing platforms, and resources that empower individuals to drive change in their own lives and communities. Organizations can also support organizational change efforts by assisting with retraining and restructuring processes.

Role of Individuals in Implementing Change:

1. Personal Choices and Conscious Consumption: Individuals can implement change by making conscious choices in their everyday lives. These choices can include adopting sustainable lifestyles, reducing personal carbon footprints, or supporting ethical businesses. Personal choices collectively have the power to influence market dynamics and drive demand for change.

2. Activism and Advocacy: Individuals can advocate for change and drive collective action by joining or supporting grassroots movements, signing petitions, participating in protests, or engaging in community organizing. Through their activism, individuals can raise awareness, bring attention to pressing issues, and influence policy decisions.

3. Volunteering and Community Engagement: Individuals can actively engage in community-level initiatives and contribute their time, skills, or resources to drive local changes. This can involve volunteering for environmental cleanup, mentoring programs, community gardens, or other social activities that promote positive change.

4. Continuous Learning and Awareness: Individuals can stay informed and educated about the pressing issues and become agents of change by sharing knowledge and influencing others. By continuously learning and staying

engaged with changes happening in various domains, individuals can inspire and motivate others to join in implementing change.

In conclusion, implementing change requires collective effort from governments, organizations, and individuals. While governments provide the policy framework, regulations, and resource allocation, organizations contribute through research, advocacy, and capacity-building. Finally, individuals play a vital role by making conscious choices, engaging in activism, volunteering, and continuously learning. Together, these stakeholders can drive meaningful and sustainable change in societies and address pressing global challenges.

Chapter 13: Future Outlook: Visions and Challenges

In this chapter, we delve into the fascinating realm of the future outlook for various industries, exploring some of the visionary ideas and challenges that lie ahead. As technology continues to advance rapidly, the possibilities seem endless, but at the same time, there are significant obstacles to overcome.

One of the most exciting areas to explore is the future of transportation. With the rise of autonomous vehicles, the potential for a revolution in how we travel is immense. Imagine a world where cars are no longer solely controlled by humans, but can navigate and interact with each other seamlessly, reducing accidents, congestion, and energy consumption. However, this vision presents significant challenges, including regulatory concerns, safety issues, and the infrastructure needed to support this new way of transportation.

Another area of interest is the future of healthcare. Just a few years ago, the idea of using AI algorithms to detect diseases or even perform surgeries seemed like science fiction. Today, it is becoming a reality. The ability to collect large amounts of data, combined with advanced machine learning algorithms, has the potential to personalize medicine and dramatically improve patient outcomes. However, protecting patient privacy and ensuring the responsible use of these technologies remain major challenges.

Likewise, as we look into the future of energy, the need for sustainable alternatives becomes increasingly pressing. The effects of climate change are becoming more evident, and the transition towards renewable energy sources is essential. Solar and wind power, as well as hydrogen fuel cells, hold tremendous promise. However, the scalability, affordability, and integration of these technologies into existing infrastructure pose significant challenges that need to be addressed for a successful energy transition.

Education is also undergoing a transformation in the face of technological advancements. Online learning platforms, virtual reality classrooms, and personalized learning algorithms offer new opportunities for students to acquire knowledge and skills. However, ensuring equitable access to quality education and addressing the digital divide are challenges that need to be overcome to fully realize the potential of technology in education.

The future of work is another important topic that warrants attention. Automation and AI have already started transforming industries, and it is predicted that many jobs will be replaced by machines. This creates the urgent need for reskilling and retraining programs to ensure the workforce remains relevant and adaptable in the face of these changes. It also raises questions about social and economic implications, such as income inequality and the distribution of wealth.

As exciting as these visions for the future are, there are also ethical considerations that cannot be ignored. The advent of technologies like artificial intelligence and genetic editing brings with them a host of ethical questions regarding privacy, equity, and the long-term consequences on society. Striking the right balance between technological progress and ethical responsibility will be a significant challenge that needs to be carefully navigated.

In conclusion, as we peer into the future, incredibly exciting prospects and challenges lie ahead. From revolutionary advancements in transportation, healthcare, and energy to transforming education and work, the potential for a better future is within our reach. However, balancing the wide-ranging impacts of these transformations with ethical responsibilities will shape the future we ultimately create. Stepping into this future with cautious optimism, we must strive to harness the power of technology for the benefit of humanity as a whole.

- Exploration of potential scenarios and projections for the future

Exploration of Potential Scenarios and Projections for the Future

In a rapidly changing world, it is essential to explore potential scenarios and projections for the future. From technological advancements to socio-political shifts, various factors will inevitably shape the world we live in. By delving into these scenarios, we can better prepare ourselves for the challenges and opportunities that lie ahead.

One scenario that captures the imagination is the meteoric rise of artificial intelligence (AI). With ongoing research and development, it is quite probable that AI will become increasingly capable in the coming decades. This evolution has the potential to revolutionize numerous industries, such as healthcare, transportation, and finance. However, it also brings forth ethical dilemmas and job displacement concerns, as AI threatens many traditional occupations.

Another prominent scenario revolves around climate change and its impact on the environment. With rising global temperatures, extreme weather events, and melting ice caps, the need to address the urgent challenges posed by climate change is becoming more apparent. Projections indicate that without substantial changes in our consumption habits and environmental policies, we may face catastrophic consequences. Therefore, exploring alternative energy sources, sustainable practices, and proactive climate policies becomes imperative for a sustainable future.

Furthermore, geopolitical shifts are profoundly shaping our world. The rise of emerging economies, such as China and India, challenges the traditional dominance of Western powers. As economic powerhouses grow, geopolitical dynamics are sure to change. This presents the need to explore potential scenarios of new global power structures, changes in alliances, and impacts on international relations.

Beyond these grand-scale scenarios, smaller yet noteworthy projections exist. Advances in medical science, for instance, give rise to the potential for extended lifespans, the eradication of certain diseases, and enhanced quality of life. However, such breakthroughs could also trigger ethical debates regarding the limitations of life extension and equitable access to healthcare.

Technological advancements also open up possibilities for space exploration, with missions increasingly venturing into the depths of our solar system and beyond. As private industries invest in space tourism and colonization, the scenario grows more feasible that humans, within the not-so-distant future, may exist beyond our planet Earth.

While these scenarios are fascinating, it is essential to remember that they are projections, based on current trends and knowledge. They allow us to anticipate potential pathways the world may follow, but they are not certainties. The future is intrinsically uncertain and subject to countless variables, making it all the more necessary for us to remain adaptable and open to change.

Nevertheless, exploring potential scenarios for the future equips us with crucial insights, which can guide decision-making at individual, organizational, and governmental levels. It enables us to identify emerging trends, develop viable strategies, and create contingency plans. By being proactive and engaging with various scenarios, we heighten our ability to shape the future rather than merely react to it.

In summary, exploration of potential scenarios and projections for the future is a valuable exercise to anticipate the challenges and opportunities ahead. From the rise of AI to climate change and geopolitical shifts, many factors demand our attention. By delving into these scenarios, we equip ourselves with valuable foresight, contributing to our collective ability to adapt, innovate, and shape a better future.

- Identification of challenges in transitioning away from fossil fuels

Transitioning away from fossil fuels is a complex and multifaceted process. As the world grapples with the urgent need to mitigate climate change and reduce greenhouse gas emissions, identifying and addressing the challenges associated with this transition becomes paramount. The following are some of the key challenges that need to be overcome as we seek to move towards a more sustainable and clean energy future.

1. Economic costs: One of the significant challenges in transitioning away from fossil fuels is the high economic cost associated with it. Fossil fuel-dependent industries, such as coal, oil, and gas, have been essential drivers of economic growth for many countries. Replacing these industries with renewable energy sources and investing in the necessary infrastructures can require substantial upfront capital investment. This financial burden often deters policymakers and decision-makers from taking bold steps towards transitioning to cleaner alternatives.

2. Existing energy infrastructure: The world's energy infrastructure has been built around fossil fuels for many decades. Power plants, pipelines, refineries, and distribution networks are designed to handle conventional energy sources and maintain the status quo. Reconfiguring and retrofitting these systems to accommodate renewable energy is not a straightforward task. Developing and implementing the necessary infrastructure to support the transition requires extensive planning, coordination, and, in some cases, disrupting existing systems. This physical infrastructure challenge poses a significant barrier to rapid and widespread deployment of cleaner energy alternatives.

3. Variable nature of renewable energy: Unlike fossil fuels, renewable energy sources like solar and wind are inherently intermittent and variable. Their availability is subject to factors such as weather conditions and geographical limitations. This variability introduces complexity into ensuring a reliable and consistent energy supply. Balancing the intermittent nature of

renewables with a baseline energy supply becomes a critical challenge, especially as the share of renewable energy in the power generation mix increases. Developing and integrating advanced storage technologies, grid management systems, and demand-response mechanisms are vital to overcome this challenge.

4. Political and legal barriers: The transition away from fossil fuels often encounters political and legal obstacles. Sectors that have significant political influence, such as the fossil fuel industry, may resist the transition to protect their economic interests. Lobbying efforts, campaign contributions, and resistance from entrenched stakeholders can hinder policymakers from implementing necessary emission reduction strategies and renewable energy targets. In some cases, legal frameworks may also be oriented towards supporting fossil fuel industries, making it difficult to enact changes favorable to cleaner alternatives.

5. Workforce and employment: The transition away from fossil fuels will undoubtedly lead to the reshaping of the job market. Industries heavily dependent on fossil fuels, such as coal mining and oil extraction, will likely see a decline in employment opportunities. Moreover, renewable energy sectors, although growing, may not immediately offer the same number of jobs as fossil fuel industries. The challenge lies in managing this transition and ensuring that, through upskilling, reskilling, and job creation in other sectors, a just and fair transition takes place, protecting individuals and communities dependent on fossil fuel-related jobs.

6. Geopolitical implications: The global reliance on fossil fuels has significant geopolitical implications. Countries with abundant oil and natural gas reserves often wield considerable economic and political influence. Shifting away from fossil fuels may upset the balance of power, disrupt global alliances, and cause political instability in some regions. This transition will require careful diplomacy and collaboration between nations to minimize potential geopolitical conflicts that may arise from the shifting energy landscape.

Despite these challenges, transitioning away from fossil fuels is imperative to mitigate the impacts of climate change and foster a sustainable future. Tackling these obstacles will require concerted efforts from governments, businesses, communities, and individuals to overcome economic, infrastructural, technological, political, workforce, and geopolitical challenges.

By acknowledging and addressing these challenges, we can pave the way for a cleaner and prosperous world powered by renewable energy sources.

92

Chapter 14: Case Studies of Fossil Fuel Transition

Abstract: In order to understand the complexities and challenges associated with the transition from fossil fuel-based energy systems to more sustainable alternatives, this chapter presents a range of detailed case studies. These case studies provide valuable insights into the various factors that influence the success or failure of such transitions, and offer important lessons for policymakers, industry stakeholders, and the wider public.

1. The Need for Fossil Fuel Transition

This section briefly discusses the urgency and necessity of transitioning away from fossil fuel-dependent energy systems to mitigate climate change, reduce environmental degradation, and improve human health. It emphasizes the importance of studying real-world examples to inform successful transition strategies.

2. Case Study 1: Germany's Energiewende

Germany's Energiewende (Energy Transition) is arguably the most well-known and scrutinized case study of a country attempting to shift from fossil fuels to renewable energy. This case study explores the origins, goals, and progress of Germany's transition, highlighting its successes, challenges, and lessons learned.

3. Case Study 2: The United States' Clean Power Plan

The United States' Clean Power Plan, introduced during the Obama administration, aimed to curb carbon emissions from power plants and encourage a transition towards cleaner energy sources. This case study delves into the development, implementation, and subsequent challenges faced by the Clean Power Plan, shedding light on the political, economic, and social factors that influenced its outcomes.

4. Case Study 3: Denmark's Experience with Wind Power

Denmark has achieved remarkable success in integrating wind power into its energy system. This case study examines Denmark's journey towards becoming a global leader in the wind energy sector, focusing on the policy frameworks, public support, and technological advancements that enabled its significant transition.

5. Case Study 4: The United Arab Emirates' Solar Energy Push

Traditionally heavily reliant on oil and gas, the United Arab Emirates has embarked on an ambitious journey to diversify its energy sources through solar power. This case study investigates the UAE's solar energy initiatives, including massive solar farms and investment in innovative solar technologies, while also considering the unique challenges faced by a resource-rich country in transitioning away from its primary energy export.

6. Case Study 5: Australia's Approach to Coal Phase-Out

Australia, one of the largest coal exporters globally, grapples with the challenge of transitioning to a low-carbon economy. This case study analyzes Australia's efforts to phase out coal, examining policies, community resistance, economic consequences, and the role of international pressure. It provides critical insights into the complexities of transitioning for countries heavily dependent on fossil fuel industries.

7. Conclusion and Key Findings

This concluding section synthesizes key findings and common themes across the presented case studies. It emphasizes the importance of tailored solutions, robust policy frameworks, public engagement, technological innovation, and international collaboration in achieving successful fossil fuel transition.

Overall, this chapter presents a comprehensive examination of real-world attempts at transitioning away from fossil fuels towards sustainable energy systems. It offers in-depth analyses, data-driven insights, and valuable lessons crucial for policymakers, researchers, and individuals seeking to better understand and navigate the complexities of achieving a sustainable energy future.

- Examination of countries and regions successfully transitioning to sustainable energy

The transition to sustainable energy is a pressing global issue as traditional forms of energy, such as fossil fuels, continue to deplete natural resources and contribute to environmental degradation. Many countries and regions are already taking proactive measures to shift towards sustainable energy sources, such as solar, wind, and hydropower. While the approach and progress may vary, examining the experiences of countries successfully transitioning to sustainable energy can provide valuable insights on effective strategies and best practices.

One such example is Denmark, often hailed as a world leader in sustainable energy. With a commitment to reducing greenhouse gas emissions and becoming entirely carbon-neutral by 2050, the country has made substantial strides in this transition. Denmark heavily invested in wind energy and consequently became a major producer and exporter of wind turbines. By 2020, roughly 50% of the country's energy consumption came from wind power, and it is expected to reach 100% by 2035. Denmark's success can be attributed to the implementation of supportive policies, strong political will, investment in research and development, and engaging citizens in the transition process.

Similarly, Costa Rica has emerged as a remarkable case study in sustainable energy transformation. The small Central American nation has set ambitious goals to achieve carbon neutrality by 2050. Costa Rica's energy sector draws primarily from renewable sources, particularly hydropower, geothermal, and solar energy. In 2015, the country ran entirely on renewable energy for a period of 299 days. Costa Rica's achievement can be credited to long-term committed institutions, robust legal frameworks, and a focus on education and awareness among its citizens.

Germany is another country that has made substantial progress in transitioning to sustainable energy. Under its Energiewende (energy transition) policy, Germany aims to phase out nuclear power by 2022 and significantly reduce greenhouse gas emissions. The country has invested heavily in solar and wind energy, making substantial contributions to the country's energy mix. Germany's transition has not been without challenges, including the need for back-up power sources during times of low renewable energy production. However, the country's commitment to decentralized energy production, community ownership of renewable installations, and comprehensive legislation have played pivotal roles in Germany's sustainability journey.

Apart from countries, certain regions have also demonstrated successful transitions to sustainable energy. For instance, the Shetland Islands in Scotland have achieved 100% renewable energy in their electricity grid through a combination of wind, tidal, and solar power. The community-led initiative exemplifies the potential of small, local energy systems in making sustainable energy accessible and affordable.

Examining these successful cases of transitioning to sustainable energy reveals some recurring themes. These include the adoption of supportive policies and legal frameworks, strong political will, investment in research and development, public engagement, and a focus on education and awareness. Additionally, successful countries and regions often undergo a systematic phasing out of traditional energy sources while simultaneously investing in diverse and decentralized renewable energy systems.

However, it is important to note that each transition is context-specific, and there is no one-size-fits-all approach. Different countries and regions have varying resources, geographical features, and technological capabilities that influence their energy transition pathways. For instance, regions with abundant natural resources like wind or solar potential may focus more on harnessing those resources, while others may prioritize a combination of different technologies.

In conclusion, the examination of countries and regions successfully transitioning to sustainable energy provides inspiring examples for the global community. Denmark, Costa Rica, Germany, and the Shetland Islands have all demonstrated the potential and benefits of shifting to renewable energy sources. Their experiences highlight the importance of supportive policies, strong political commitment, public engagement, and a holistic approach to transitioning from traditional to sustainable energy systems. By studying these successful experiences, policymakers and stakeholders can learn valuable lessons to inform their own sustainable energy transition strategies and accelerate progress towards a more sustainable future.

- Analysis of challenges faced and lessons learned

In today's rapidly changing business landscape, organizations are constantly faced with various challenges that can hinder their growth and success. These challenges can come from different sources such as market conditions, organizational structure, technology, or competition. In order to navigate through these challenges, organizations need to understand the lessons learned from past experiences and adapt their strategies accordingly.

One of the major challenges faced by organizations is the ever-increasing competition. With globalization, companies now have to compete not only with local players but also with international giants. The lesson learned here is the importance of constant innovation and staying ahead of the curve. Organizations must continuously strive to develop new and improved products and services to stay competitive. They must also invest in research and development to anticipate market trends and even disrupt their own industry before someone else does.

Another common challenge organizations face is technological disruption. The advancement of technology has become a double-edged sword for many companies. On one hand, it offers tremendous opportunities for efficiency, automation, and cost reduction. On the other hand, it can render existing business models and processes obsolete. The lesson learned here is the need for organizations to proactively embrace and adapt to new technologies. They must invest in digital transformation initiatives to stay relevant and exploit the benefits of emerging technologies such as artificial intelligence, data analytics, and internet of things.

Organizational structure and culture can also pose challenges to organizations. A hierarchical and bureaucratic structure can hinder communication, innovation, and agility. The lesson learned here is the need for organizations to adopt a more flat and agile structure. They must empower employees at all levels to make decisions and collaborate across functions and departments. Cultural barriers must be removed to foster open communication and a culture of continuous learning and improvement.

Market conditions and economic volatility present challenges to organizations as well. Market demand can fluctuate, and economic downturns can impact consumer spending. The lesson learned here is the importance of diversification and agility. Organizations must not rely solely on one product or market. Instead, they should diversify their portfolio and expand into new markets or industries. They must also build resilience in their operations, supply chain, and financials to weather economic uncertainties.

Lastly, one of the major challenges faced by organizations is the effective management of human resources. Recruiting, engaging, and retaining top talent is crucial for success. The lesson learned here is the need for organizations to prioritize employee well-being, professional development, and recognition. Creating a positive work environment, providing competitive compensation and benefits, and offering opportunities for growth and advancement are key strategies to attract and retain top talent.

In conclusion, organizations face various challenges in today's business landscape. However, by understanding the lessons learned from past experiences, they can adapt their strategies and overcome these challenges. Continuous innovation, embracing technology, adopting agile structures, diversifying markets, and prioritizing employee well-being are some of the key lessons to be considered for successful navigation through these challenges.

Chapter 15: Continued Role of Fossil Fuels in the Global Market

Fossil fuels have played a significant role in the global market for over a century. They have been the backbone of industrial revolutions and economic growth worldwide. Despite growing concerns about environmental sustainability and the push for renewable energy sources, fossil fuels continue to dominate the energy landscape. This chapter examines the reasons behind the continued role of fossil fuels in the global market, exploring various aspects such as economic factors, geopolitical dynamics, and technological limitations.

1. Economic factors:

One of the primary reasons for the enduring dominance of fossil fuels is their cost-effectiveness. Coal, natural gas, and oil have high energy density and are economically viable in large-scale energy production. Fossil fuel extraction and processing infrastructures have been developed over the years, making the industry highly efficient and cost-competitive. Additionally, the abundant global reserves of fossil fuels ensure a steady supply, reducing the risk of price volatility.

Moreover, the fossil fuel industry has established a vast network of transportation, storage, and distribution systems. This allows for easy accessibility and trade on an international scale. These economic factors contribute to the attractiveness of fossil fuels for both energy producers and consumers.

2. Geopolitical dynamics:

Fossil fuel resources are not distributed evenly across the globe. Certain regions, such as the Middle East and Russia, possess vast reserves of oil and natural gas, giving them significant geopolitical influence. The reliance of many nations on these regions for their energy needs creates interdependencies and geopolitical complexities. This has resulted in tense relationships and conflicts, such as those triggered by attempts to control oil-producing regions or pipelines. Governments often act to safeguard their fossil fuel interests, further entrenching the importance of these resources in global politics and the global market.

3. Technological limitations:

While renewable energy sources have made impressive advances in recent years, there are still technological limitations and infrastructure challenges to overcome. Fossil fuel power plants and engines are well-established and reliable, requiring minimal retrofitting or replacement for continued use. Conversely, renewable energy technologies often involve significant upfront costs and require substantial investment in infrastructure development and grid integration. As a result, many countries with limited financial resources or less technologically developed infrastructures find it difficult to transition away from fossil fuels.

Beyond energy production, fossil fuel derivatives are key ingredients in numerous industries, including petrochemicals, plastics, and pharmaceuticals. These industries rely heavily on fossil fuels, as alternatives are often less cost-effective or challenging to develop. As such, the dominance of fossil fuels goes beyond energy consumption alone.

4. Government policies and subsidies:

Government policies and subsidies also contribute to the continued role of fossil fuels in the global market. Some nations support their fossil fuel industries through subsidies and tax incentives, making them more economically viable compared to renewable energy options. This has slowed down the transition to cleaner energy sources and reinforced the dominance of fossil fuels. However, it is essential to note that in recent years, there has been a global push toward reducing fossil fuel subsidies and promoting renewables, though progress remains slow.

DESPITE THE GROWING global awareness of the negative environmental impacts of fossil fuel consumption, the continued role of fossil fuels in the global market is multi-faceted. Economic factors, geopolitical dynamics, technological limitations, and government policies all contribute to the resilience of fossil fuels. While renewable energy sources are gaining momentum, they still face challenges in achieving cost-competitiveness, scalability, and infrastructure development. Overcoming these hurdles and transitioning to cleaner energy systems remains a pressing global goal, but it is clear that the fossil fuel industry will continue to play a significant role in the global market for the foreseeable future.

- Discussion on the perceived necessity of fossil fuels in certain industries

Perceived Necessity of Fossil Fuels in Certain Industries: A Discussion on their Longevity and Alternatives

FOSSIL FUELS HAVE LONG been the lifeblood of various industries, powering sectors such as transportation, manufacturing, and energy production. However, discussions surrounding their perceived necessity in society have become increasingly prevalent in recent years. This article delves into the intricate details, providing an insightful exploration of the factors that contribute to their continued prominence while examining potential alternatives for a sustainable future.

Historical Significance and Industrial Dependency:

From air and sea travel to heavy construction machinery, fossil fuels have played a vital role in modern manufacturing, enabling great advancements and shaping our developed societies. Their energy density, readily available infrastructure, and widespread production capabilities have resulted in a high level of industrial dependence.

Transportation has long been one of the sectors most reliant on fossil fuels, due to the limited range and energy storage capacity of alternative options. In this aspect, gasoline and diesel continue to dominate, owing to their ease of use, portability, and well-established refueling infrastructure. Additionally, other energy-dense fossil fuels such as jet fuel continue to be essential for air travel, further solidifying the perceived necessity of these energy sources within the transportation industry.

Energy Production Challenges and the Role of Fossil Fuels:

When considering global electricity generation, fossil fuels remain a dominant force. Power plants reliant on coal or natural gas provide a stabilizing base of energy, facilitating instant responsiveness to fluctuations in power demand. Non-renewable energy sources contribute nearly 64% of the world's total electricity generation.

Renewable alternatives, such as solar and wind, show immense potential for future energy generation. However, challenges regarding their intermittent availability, energy storage capacity, and infrastructure integration hinder their broad-scale adoption. As such, fossil fuel-powered plants continue to be viewed as necessary to maintain stability and supplement renewable energy sources, at least for the time being.

Economic Implications and Technological Advancements:

Historically, fossil fuels have acted as catalysts for economic growth by providing affordable and reliable energy. Switching to alternative energy sources, on the other hand, poses certain economic challenges. The extraction, refinement, distribution, and utilization of alternative energies entail significant upfront investments and often have lower production capacity, posing potential financial hurdles for current industries.

Nonetheless, rapid technological advancements and the continually decreasing costs of renewable energy sources have started gaining traction, making them increasingly competitive economically. Countries are steadily transitioning to cleaner alternatives, backed by policies promoting sustainability and climate goals. These trends reflect an evolving perception of the necessity of fossil fuels.

WHILE THE PERCEIVED necessity of fossil fuels in certain industries remains prevalent today, numerous factors are influencing the growing discussion about their continued reliance. The expansion of renewable energy technologies, socio-political commitments toward environmental sustainability, and economic shifts are progressively driving the transition towards cleaner alternatives. As technology continues to advance, it is likely that alternative energy sources will become increasingly attractive and viable, shifting the dialogue around the necessity of fossil fuels altogether.

Chapter 16: Addressing the Fossil Fuel Industry's Responsibility

The fossil fuel industry has played a significant role in powering our world for decades. However, in recent years, the environmental impacts and potential detriments associated with fossil fuels have become apparent. As concerns about climate change and air pollution intensify, it is essential to address the industry's responsibility in mitigating and rectifying these issues. This chapter delves into the complex arena of the fossil fuel industry's responsibility and provides detailed analysis and discussion regarding various solutions and approaches.

1. Historical Context:

To understand the current state of affairs, it is crucial to examine the historical context of the fossil fuel industry. The rise of coal, followed by oil and natural gas, transformed the world's energy consumption patterns. Economic growth, technological advancements, and geopolitical dynamics have all been deeply intertwined with the fossil fuel industry. Recognizing this interconnectedness helps us better grasp the challenges, both practical and political, of addressing the industry's responsibility.

2. Environmental Impact:

The impact of fossil fuels on the environment is undeniable. Greenhouse gas emissions from burning fossil fuels are the leading contributors to climate change. The extraction and combustion of fossil fuels also contribute to air and water pollution, biodiversity loss, habitat destruction, and environmental degradation. Addressing the fossil fuel industry's responsibility necessitates a holistic approach that combines reducing emissions, promoting cleaner technologies, and undertaking environmental conservation efforts.

3. Corporate Accountability:

One key aspect of addressing the fossil fuel industry's responsibility is ensuring corporate accountability. Major companies within the industry have often come under scrutiny for their actions and lack of transparency. This chapter explores various mechanisms to hold fossil fuel companies accountable, including shareholder activism, divestment campaigns, and legal avenues. It also highlights the responsibility governments have in regulating and enforcing environmental safeguards within the industry.

4. Transition to Clean Energy:

Transitioning away from fossil fuels towards renewable and clean energy sources is an integral part of addressing the industry's responsibility. This chapter examines the potential policy measures, incentives, and investments required to facilitate this transition. It explores the role of governments, businesses, and individuals in driving innovation and creating a sustainable energy future.

5. Just Transition:

A just transition is a crucial element in addressing the fossil fuel industry's responsibility. This concept revolves around mitigating the impacts of transitioning away from fossil fuels on workers, communities, and economies that heavily rely on the industry. The chapter delves into the social, economic, and political considerations necessary to ensure a fair and equitable transition that does not leave vulnerable populations behind.

6. International Cooperation:

Addressing the fossil fuel industry's responsibility is an inherently global challenge that necessitates international cooperation. This chapter explores avenues for collaboration among nations, organizations, and stakeholders in transitioning to cleaner energy systems. It emphasizes the importance of shared responsibility and collective action in combating climate change and reducing reliance on fossil fuels.

THE FOSSIL FUEL INDUSTRY'S responsibility in mitigating environmental damage, promoting sustainable practices, and fostering a clean energy transition is critical. This chapter serves as a comprehensive resource providing detailed analysis, insights, and recommendations for addressing these responsibilities. By acknowledging the complexity of the issue and engaging in meaningful discussions, we can pave the way for a more sustainable future.

- Examination of ethical considerations and the industry's accountability

Examination of ethical considerations and the industry's accountability in today's world is of utmost importance. With the rapid advancements in technology and globalization, businesses have faced unprecedented challenges regarding ethics and accountability. It is crucial to scrutinize the ethical practices of industries to ensure that they are accountable for their actions and uphold ethical standards in today's complex and interconnected world.

Ethical considerations include a range of factors, such as transparency, fairness, respect for human rights, environmental sustainability, and social responsibility. Each industry has unique ethical challenges it must address to maintain accountability. For example, the manufacturing industry faces ethical dilemmas related to worker safety, product quality, and the impact of their operations on the environment. The healthcare industry grapples with issues surrounding patient privacy and the accessibility and affordability of healthcare services.

In recent years, there have been numerous high-profile cases where industries have failed to uphold ethical standards, leading to severe consequences. From the Enron scandal to Volkswagen's emissions cheating scandal, these incidents have highlighted the need to examine the industry's accountability rigorously. Such cases have profound effects not only on the companies involved but also on trust in the industry as a whole.

One way to ensure industry accountability is by implementing stricter regulations and laws. Government bodies play a crucial role in monitoring and enforcing ethical standards, making businesses legally accountable for their actions. For instance, the Sarbanes-Oxley Act in the United States was introduced to hold accountable those in the accounting industry responsible for financial reporting fraud. Similarly, the General Data Protection Regulation (GDPR) in the European Union holds companies in the technology and data-driven industries accountable for protecting individuals' privacy.

Moreover, companies themselves must take a proactive approach to accountability and ethics. Corporate social responsibility (CSR) initiatives demonstrate a company's commitment to ethical practices beyond what is legally mandated. CSR includes initiatives such as fair trade, philanthropy, and sustainable business practices. By voluntarily adopting such practices, companies establish their accountability and contribute positively to society.

Another essential aspect of industry accountability is consumer empowerment. In today's digital age, consumers have become more conscious of ethical considerations and demand transparency and accountability from the companies they engage with. They influence industries through their purchasing decisions, public opinion, and social media movements. Therefore, companies are incentivized to be more ethical and accountable to maintain their competitive edge and reputation.

The examination of ethical considerations is vital as it promotes responsible business practices, fosters trust in industries, and contributes to the greater good of society. Ethical companies ensure the fair treatment of their employees, actively work toward minimizing their environmental impact, and take steps to address any societal issues caused by their operations.

In conclusion, ethical considerations and industry accountability are crucial components of our modern world. Governments, companies, and consumers must collaborate to ensure industries are held accountable for their actions and operate in an ethical manner. By doing so, we can create a sustainable and responsible business environment that benefits both organizations and society as a whole.

- Strategies for holding corporations accountable and pushing for change

Strategies for holding corporations accountable and pushing for change are crucial in today's complex global market. With many corporations wielding immense power and influence over various aspects of our lives, it is essential for individuals, organizations, and governments to ensure that these entities are held responsible for their actions. Here are some strategies that can be employed to achieve this goal:

1. Legal Mechanisms: One of the primary ways to hold corporations accountable is through legal mechanisms, such as legislation and regulation. Governments at the local, national, and international levels can pass laws that address corporate behavior, ensure transparency, and promote accountability. These regulations can include measures to protect the environment, labor rights, consumer rights, and prevent unethical practices.

2. Promote Stakeholder Engagement: Corporations have a range of stakeholders, including employees, customers, shareholders, and the communities in which they operate. Empowering and engaging these stakeholders can be an effective strategy to hold corporations accountable. Stakeholders should have the opportunity to voice their concerns through increased transparency, access to information, and channels for feedback. Activist groups can organize campaigns to raise awareness and mobilize stakeholders to demand change.

3. Strengthen Corporate Social Responsibility (CSR): Encouraging corporations to adopt comprehensive CSR practices can contribute to holding them accountable. CSR initiatives often go beyond legal requirements and involve corporations voluntarily taking responsibility for their impact on society and the environment. By promoting transparency, ethical practices, and proactive efforts to address social and environmental issues, corporations can demonstrate their commitment to accountability.

4. Active Shareholder Advocacy: Shareholders of companies can play a critical role in pushing for change. By leveraging the power of their investments,

shareholders can engage in proxy voting, dialogue with management, and file shareholder resolutions to hold corporations accountable. Shareholders can demand greater transparency, responsible corporate governance, and actions to mitigate negative impacts on society and the environment.

5. Consumer Awareness and Activism: Informed and conscious consumer choices can be a powerful tool for change. By supporting ethically responsible companies and exercising their buying power, consumers can send strong market signals to corporations. Consumer activism, including boycotts and social media campaigns, can also create pressure on corporations to change unethical practices.

6. Media and Public Pressure: The media plays a crucial role in exposing corporate wrongdoings and raising public awareness. Investigative journalism can shed light on unethical practices, corporate misconduct, and environmental or social impacts. Public pressure, through awareness campaigns, social media advocacy, and public protests, can amplify demands for change and hold corporations accountable.

7. International Collaboration: Many corporations operate globally, making it imperative to address their accountability on an international scale. International collaboration among governments, civil society organizations, and other stakeholders is vital. By sharing best practices, aligning standards, and collectively monitoring corporations' actions, international efforts can push for responsible corporate behavior.

Holding corporations accountable and pushing for change is a multifaceted process that requires concerted efforts from various stakeholders. By employing these strategies, individuals, organizations, and governments can work together to create a more responsible corporate landscape that ensures the well-being of people and the planet.

- Final thoughts on the relationship between the fossil fuel industry and global warming

The relationship between the fossil fuel industry and global warming is a complex and contentious issue that has garnered significant attention in recent years. Fossil fuels, including coal, oil, and natural gas, have long been the dominant source of energy worldwide. However, the burning of these fuels releases large quantities of greenhouse gases, primarily carbon dioxide, into the atmosphere, leading to a steady rise in global temperatures.

Numerous scientific studies have established that human activities, particularly the burning of fossil fuels, are the primary driver of global warming. These emissions trap heat in the Earth's atmosphere, causing the planet to warm rapidly. The consequences of this warming are already becoming apparent, with rising sea levels, more frequent and intense extreme weather events, and environmental degradation impacting numerous communities around the world.

Given that the fossil fuel industry is at the forefront of the production and consumption of these energy sources, it bears a significant responsibility for its contribution to climate change. The industry has faced criticism for its role in perpetuating an unsustainable energy system and for actively obstructing climate action. This has been exemplified by efforts to cast doubt on established scientific consensus, funding climate change denial campaigns, and lobbying against regulations and policies aimed at reducing greenhouse gas emissions.

The global warming-fossil fuel relationship is also tainted by the industry's disproportionate influence and power over political and economic systems. Lobbying and campaign contributions have allowed fossil fuel companies to exert considerable control and influence over governments, resulting in a lack of decisive action to address climate change. This has often led to policies that favor the continuation of fossil fuel use and hinder the uptake of clean and renewable alternatives.

On the other hand, the fossil fuel industry argues that it plays a crucial role in supporting global economic development and meeting the energy demands of a growing global population. They contend that transitioning away from fossil fuels is a complex and challenging task that requires time, research, and substantial investment. They argue that a sudden abandonment of fossil fuels would have severe economic consequences, including job losses and energy shortages.

Navigating this complex relationship between the fossil fuel industry and global warming requires a multi-faceted approach. It is increasingly clear that the world needs to shift towards cleaner and renewable sources of energy to mitigate the impacts of climate change. Innovative technologies such as wind, solar, and geothermal offer promising alternatives that can help reduce reliance on fossil fuels. However, transitioning away from fossil fuels must also take into account the socioeconomic challenges and the need for a just transition for impacted communities.

In conclusion, the relationship between the fossil fuel industry and global warming is a contentious and challenging one. The industry plays a pivotal role in generating the energy that drives economic growth and development globally but is also a significant contributor to greenhouse gas emissions. Addressing the global warming crisis requires holding the fossil fuel industry accountable for its environmental impact while actively seeking collaboration and solutions that balance energy needs with sustainable and responsible practices. This includes transitioning to cleaner sources of energy, supporting affected communities, and advocating for strong climate policies that prioritize the environment and future generations.